Contents

KU-007-295

Preface

Assessing the construction cost of a building project is a continuous process that commences when the client asks his professional advisors what it will cost. It ends when the cheque for the release of the last portion of the retention is paid and the final account is settled.

Whether a project is profitable or not can depend on decisions taken at the first stages of its life. The people who carry out this assessment bear the responsibility of deciding whether the cost information available at the time warrants spending further time and money investigating the viability of a project that is usually still at the conceptual stage.

It is vital, therefore, that the most accurate methods of calculating the probable costs of construction are used and I hope that the information in this book will help clients, developers, architects, engineers and surveyors involved in this process.

Time spent on the appraisal of projects that do not proceed is not always wasted – lessons can be learnt that will help the next project. The greater waste lies in allowing jobs to proceed without the people making the key decisions having the best cost information available to them.

It is not possible to provide cost data in the early stages of a job that is totally accurate, only a historical analysis of the final account can do that, but guidelines can be set out that should allow the client commissioning the work to help make the right decision before becoming contractually committed to spending large sums of money on land purchase, professional fees and construction costs.

On a well-managed project, the cost plan will be monitored continuously as changes are made to the original design or for other reasons. If the first stage budget is based on reliable cost data, the project should have a better chance of being completed within budget.

I have received a great deal of help in the preparation of this book and would like to thank those suppliers and contractors who gave me their time and support. I am indebted to John McGee for his helpful comments in general and his work on the Life Cycle Costing section in particular.

Although every effort has been made to ensure the accuracy of the information, neither the Publishers nor I can accept any type of liability resulting from the use of the contents.

On a more positive note, I would welcome constructive criticism of ways to make the contents more relevant to the changing needs of the construction industry for future editions.

Bryan Spain
June 2000

Introduction

This edition of *Spon's First Stage Estimating Handbook* is aimed at all members of the construction team who have an involvement in construction costs during the early stages of the development of a project. There are many factors that can affect the profitability of a job over which the construction team has little or no control.

It is important, therefore, that areas that can be influenced by good management skills should be given special attention. This applies particularly to the first stage estimating process when a solid base of the project's financial position can be established.

The contents have been laid out in the order that most projects are financially assessed. The quality of the cost information, however, is almost wholly dependent upon the quality of the specification and design information available at that stage of the appraisal.

Chapter 1 lists square metre rates for building and M&E work for a wide variety of building types. The rates are expressed as a range based on historical costs and should provide a client with a broad indication of likely costs.

Chapter 2 fulfils the same function but the costs are given in unit costs, e.g. the cost per cinema seat or per cover in a restaurant. This method of assessment can be surprisingly accurate but should always be backed up by more detailed methods of cost appraisal.

Chapter 3 contains elemental cost analyses of 32 different types of buildings. These analyses provide data on the percentage and cost breakdown of 24 elements for each building and are a useful tool in identifying imbalances between the elements of different buildings.

Chapter 4 includes unit rates for building, landscaping, civil engineering, M&E and alteration work. Where possible composite rates are displayed which combine several different descriptions to provide a single rate for separate but linked activities. For example, excavation, concrete and brickwork up to DPC level are combined to produce a single rate for strip foundation.

The use of composite rates can save valuable time in the preparation of the cost plan. Where item descriptions do not lend themselves to be combined with others they are listed as principle rates.

Chapter 5 provides indices reflecting historical costs of construction costs and tender prices. **Chapter 6** deals with property insurance and contains an example of re-building costs for insurance purposes. **Chapter 7** sets out the cost of employing professional advisors whose fees are now negotiable and not mandatory.

Chapter 8 contains information on life cycle costing which is now recognised as an important factor in the assessment of overall construction costs. **Chapters 9** and **10** contains information that may be useful when preparing a first stage estimate.

At this stage of a project the information available is limited and although the brief from a client can be minimal an approximation of costs must still be prepared. Here is a simple example.

It is proposed to construct a single story advance factory of 8,000 square metres in the South West of England. An outline plan has been prepared showing parking areas and a road layout. It is known that local contractors are operating at full capacity so tender levels are high. The work would commence in January 2001 with a nine month construction period.

Main building   £

8000 m2 at £80 (Chapter 1)   640,000

External works
(based on rough quantities and rates in Chapter 4)

Car park	70,000	
Roads	90,000	
Fencing	15,000	
Drains	22,000	
Sort landscaping	10,000	
Lighting	8,000	215,000

Carried forward   855,000

	£
Brought forward	855,000
Extra for anticipated high tender levels say 6%	<u>51,300</u>
	906,300
Allow for inflation say 9 months @ 4% × £906,300	<u>27,189</u>
	933,489
Allow for regional variations (see Chapter 5) say South West, 95% × £933,489	886,815
Professional fees (see chapter 5)	
say 12% × £886,815	<u>106,418</u>
	<u>993,233</u>

Depending upon the quality of the information on which the first stage estimate is based, a basement and ceiling band of between 5% and 10% should be applied and reported to the client:

5%	say	£950,000 to £1,050,000
10%	say	£890,000 to £1,090,000

These costs exclude land purchase and VAT and any special costs connected to the project. They also exclude any development grants or subsidies that may be available in the area involved.

1

Costs per square metre

BUILDING WORK

These square metre prices exclude the cost of external works, fittings, furniture and professional fees. They are stated in a range and are intended to provide a broad cost of the work.

Public service buildings	£
Ambulance stations	650-850
Banks	
local branch	950-400
city	1250-1800
Building societies	
local branch	900-1350
city	1200-1650
Fire stations	875-950
Magistrates courts	950-1250
County courts	1150-1350
Police stations	750-1050
Prisons	950-1250
Post offices	850-1000
Public conveniences	900-1150

	£
Town halls	1000-1250
Village halls	620-750

Industrial buildings

Agricultural
     livestock   200-275
     storage   150-200

Factories
     light industrial   325-450
     heavy industrial   425-650
     extra for owner occupation   100-150
     extra for office accommodation

High tech. laboratory   750-1200
     extra for air conditioning   150-225

Nursery units   300-375

Warehouses/stores, shell only
     up to 2000m2 floor area   375-500
     over 2000m2 floor area   250-400
     extra for owner occupation   250-350

Offices, business park, shell and core   475-650
     high tech. unit   650-900
     extra for air-conditioning   150-225
     extra for owner occupation   250-350

Offices, city centre, shell and core   700-850
     extra for 3 to 5 floors   200-250
     extra for 5+ floors   250-300
     extra for air conditioning   150-225
     extra for prestige building   200-350
     extra for owner occupation   250-350

Health and welfare facilities   £

Day surgeries	1000-1500
Group surgeries	700-900
Homes for the mentally handicapped	900-1150
Health centres	850-1100
Hospitals	
general	850-1150
laboratories	900-1200
pharmacies	800-1100
private	950-1500
teaching centres	750-1100
Nursing homes	750-1100

Leisure

Canteens	750-1000
Cinemas	
shell only	500-650
complete	900-1250
Community centres	750-1000
Concert halls	1300-1800
Exhibition buildings	900-1500
Ice rinks	850-1000
Golf club houses	750-1100
Motorway service buildings	900-1100
Public houses	700-825

4   Costs per square metre

	£
Restaurants	
shell only	450-600
complete	900-1400
Sports centres	600-700
Sports pavilions	550-800
Squash courts	550-750
Swimming pools	
school	950-1300
local	900-1200
international	1200-1600
fun	1300-1800
Theatres	
small	850-1200
prestige	1400-2000

Religious buildings

Churches, chapels	750-1100
Church halls, meeting houses	650-900
Convents	800-950
Crematoria	850-1250
Temples, synagogues, mosques	800-1100

Educational buildings

Colleges	650-900
Conference centres	1000-1650
Laboratories	900-1450

	£
Lecture theatres	900-1200
Libraries	
local	700-950
city centre	850-1200
Museums	
local	750-950
city centre	950-1300
Research facilities	950-1400
Schools	
nursery	650-1300
primary	600-900
secondary	600-800
special	650-950
Universities	750-1650

Residential buildings

	£
Guest houses	750-1000
Hotel refurbishment	
3 star	500-650
5 star	750-1000
Private housing	
bungalows	550-750
houses, detached	
2 bedrooms	550-600
3 bedrooms	650-700
4 bedrooms	750-800
houses, semi-detached	
2 bedrooms	550-600
3 bedrooms	600-650

	£
flats, low rise	
standard	550-650
luxury	800-950

Public housing schemes

	£
bungalows	
1 bedroom	500-550
2 bedrooms	525-575
houses	
1 bedroom	450-500
2 bedrooms	475-525
3 bedrooms	500-550
flats, low rise	525-600
sheltered housing	
single storey	750-800
two storey	750-850

	£
Student accommodation	500-650
Youth hostels	550-700

Transport

	£
Airport hangars	700-950
Airport terminals	1350-1550
Air traffic control buildings	1200-2000
Car parking	
multi-storey	250-350
underground	450-600
surface	50-75
Car showrooms	650-875
Coach and bus stations	700-900

	£
Petrol stations	750-1350
Port and harbour buildings	950-1200
Vehicle repair shops	500-750

Retail

	£
Fast food units	
shell only	400-500
complete	750-1100
Retail warehouses	
shell only	250-425
complete	550-650
Shops	
shell only	400-500
complete	700-900
Shopping centres	
standard	800-1350
malls	850-1250
Supermarkets	
shell only	400-600
complete	900-1250

MECHANICAL AND ELECTRICAL WORK

These rates per square metre are for all engineering systems of specific building types applied to the gross internal floor area of buildings excluding professional fees and VAT.

Public service buildings	**£**
Ambulance stations	130-150
Banks	130-180
Council offices	200-250
Fire stations	150-240
Magistrates courts	150-280
Police stations	130-170

Industrial buildings	
Agricultural sheds	40-60
Factories	
advance	40-90
high technology	240-400
purpose built	70-150
Warehouses	50-75
Workshops (excluding machinery)	60-140

Educational, cultural and scientific buildings	
Computer buildings	200-350
Laboratories	390-540
Libraries	180-260

	£
Primary schools	100-200
Secondary schools	100-170
Universities	130-250

Entertainment and recreation buildings

Community centres	100-180
Gymnasia	120-160
Hotels	250-350
Public houses	100-200
Restaurants	120-320
Social centres	100-180
Sports halls	130-200
Swimming pools/sport facilities	240-400

Health and welfare buildings

Children's homes	80-120
General hospitals	300-450
Health centres	140-200

Residential buildings

Local Authority and Housing Association schemes	
high rise flats	90-120
4/5 person two storey houses	60-100

10 Costs per square metre

	£
Private developments	
high rise flats	100-140
4/5 person two storey houses	70-120

2

Costs per unit

The following represent the cost per unit of a range of types of buildings. This method of assessing construction costs is usually the first step in the process of the consideration of the viability of project.

	Unit	£
Health and welfare facilities		
Hospitals (general)	Bed	50,000-85,000
Hospitals (private)	Room	60,000-140,000
Nursing homes		
old peoples	Bed	25,000-40,000
childrens	Bed	20,000-32,000
Leisure		
Cinemas	Seat	1,750-2,000
Restaurants	Cover	1,500-1,800
Sports stadium, new stands		
single tier stand	Seat	300-600
multi-tier stands with hospitality boxes	Seat	750-1,000
Hotels		
3 star	Bedroom	30,000-45,000
5 star	Bedroom	80,000-120,000

	Unit	£
Residential		
Local Authority housing	Bed	9,000-11,000
Local Authority flats	Bed	14,000-16,000
Sheltered housing	Bedroom	30,000-45,000
Private housing	Bed	11,000-14,000
Private flats	Bed	15,000-18,000
Student accommodation	Bedroom	12,000-15,000
Transport		
Car parking		
multi storey	Car	3,000-6,000
underground	Car	10,500-16,000
surface	Car	1,000-1,500
Educational buildings		
Nursery schools	Pupil	2,500-4,000
Primary schools	Pupil	3,000-5,000
Secondary schools	Pupil	4,500-5,500

3

Elemental costs

This chapter contains elemental cost breakdowns for both building and mechanical and electrical work. The building section is comprised of 33 different types of buildings in 8 categories, each building broken down into 24 cost elements. The M&E section contains elemental costs of six types of buildings.

The cost data should be used with caution because imbalances can occur in the use of elemental costs in isolation. The figures are an amalgam of the costs for a range of buildings in each category and represent a broad indication of costs rather than an accurate statement of detail.

Nevertheless, the information provided should be an invaluable tool in assessing the relative values of elements in different buildings. The tables will be particularly useful in the early cost planning process and also in the evaluation of tenders.

The costs cover building work only and excludes the costs of drainage, external works, contingency sums, professional fees and VAT.

Due to rounding off, there may be some minor discrepancies in individual element costs and totals. The figures are based upon costs prevailing in the first quarter of 2000. The following buildings are included.

BUILDING WORK

Public service buildings

Ambulance station
City bank
County court
Fire station
Police station
Village hall

Industrial buildings

Factory, light industrial
Factory, heavy industrial
Livestock building
High tech. laboratory
Nursery units
Warehouse shell
Warehouse complete

Health and welfare

Group surgery
Health centre
Old persons' nursing home
Welfare centre

Leisure

Golf club house
Public house
Restaurant
Sports hall

Education

Library
Primary school
Secondary school
Sixth form college
Special school
Teachers' training college

Residential

Local Authority low rise flats
Local Authority housing
Private flats
Luxury private flats

Transport

Multi-storey car park

MECHANICAL AND ELECTRICAL WORK

Hotel
Hospital
Leisure complex
Office block
Sports stadium
University

BUILDING WORK

Category: Public service
Type: Ambulance station
Floor area: 375m2
Total cost: £298,125

	Elemental cost £	% of cost	Cost m2 floor area £
1 Preliminaries	32,799	11	87
2 Substructure	26,831	9	72
3 Frame	-	-	-
4 Upper floors	-	-	-
5 Roof	38,756	13	100
6 Staircases	-	-	-
7 External walls	23,850	8	64
8 Windows and external doors	11,925	4	32
9 Partitions and internal walls	8,943	3	24
10 Internal doors	5,962	2	16
11 Wall finishes	5,962	2	16
12 Floor finishes	14,906	5	40
13 Ceiling finishes	2,981	1	8
14 Fittings and furnishings	20,868	7	56
15 Sanitary appliances/disposal installation	11,925	4	32
16 Hot and cold water services	5,962	2	16
17 Heating and air treatment installation	11,925	4	32
18 Ventilation installation	2,981	1	8
19 Gas services	2,981	1	8
20 Electric installation	23,850	8	64
21 Lift and conveyor installation	-	-	-
22 Protective and communication installation	5,962	2	16
23 Special installations/services equipment	11,925	4	32
24 Builders' work	26,831	9	72
Totals	298,125	100	795

Category: Public service
Type: City bank
Floor area: 907m2
Total cost: £1,173,658

	Elemental cost £	% of cost	Cost m2 floor area £
1 Preliminaries	105,640	9	116
2 Substructure	82,156	7	90
3 Frame	117,365	10	129
4 Upper floors	46,946	4	52
5 Roof	70,419	6	77
6 Staircases	23,473	2	26
7 External walls	176,048	15	194
8 Windows and external doors	58,682	5	65
9 Partitions and internal walls	35,209	3	39
10 Internal doors	23,473	2	26
11 Wall finishes	23,473	2	26
12 Floor finishes	46,946	4	52
13 Ceiling finishes	35,209	3	39
14 Fittings and furnishings	35,209	3	39
15 Sanitary appliances/disposal installation	11,736	1	13
16 Hot and cold water services	23,473	2	26
17 Heating and air treatment installation	35,209	3	39
18 Ventilation installation	11,736	1	13
19 Gas services	-	-	-
20 Electric installation	58,682	5	65
21 Lift and conveyor installation	-	-	-
22 Protective and communication installation	35,209	3	39
23 Special installations/services equipment	35,209	3	39
24 Builders' work	82,156	7	90
Totals	1,173,658	100	1,294

Category: Public service
Type: County court
Floor area: 948m2
Total cost: £1,144,236

	Elemental cost £	% of cost	Cost m2 floor area £
1 Preliminaries	114,423	10	121
2 Substructure	57,211	5	60
3 Frame	80,107	7	85
4 Upper floors	57,211	5	60
5 Roof	80,096	7	85
6 Staircases	34,327	3	36
7 External walls	102,981	9	109
8 Windows and external doors	57,211	5	60
9 Partitions and internal walls	34,327	3	36
10 Internal doors	45,769	4	48
11 Wall finishes	34,327	3	36
12 Floor finishes	45,769	4	48
13 Ceiling finishes	34,327	3	36
14 Fittings and furnishings	57,211	5	60
15 Sanitary appliances/disposal installation	22,884	2	24
16 Hot and cold water services	11,442	1	12
17 Heating and air treatment installation	22,884	2	24
18 Ventilation installation	11,442	1	12
19 Gas services	11,442	1	12
20 Electric installation	68,654	6	74
21 Lift and conveyor installation	22,884	2	24
22 Protective and communication installation	11,442	1	12
23 Special installations/services equipment	22,884	2	24
24 Builders' work	102,981	9	109
Totals	1,144,236	100	1,207

Category: Public service
Type: Fire station
Floor area: 490m2
Total cost: £456,680

	Elemental cost £	% of cost	Cost m2 floor area £
1 Preliminaries	45,668	10	93
2 Substructure	27,400	6	56
3 Frame	22,834	5	47
4 Upper floors	13,700	3	28
5 Roof	36,534	8	74
6 Staircases	9,133	2	19
7 External walls	45,668	10	93
8 Windows and external doors	27,400	6	56
9 Partitions and internal walls	13,700	3	28
10 Internal doors	13,700	3	28
11 Wall finishes	13,700	3	28
12 Floor finishes	18,267	4	37
13 Ceiling finishes	4,566	1	9
14 Fittings and furnishings	9,133	2	19
15 Sanitary appliances/disposal installation	9,133	2	19
16 Hot and cold water services	13,700	3	28
17 Heating and air treatment installation	13,700	3	28
18 Ventilation installation	4,566	1	9
19 Gas services	4,566	1	9
20 Electric installation	45,668	10	93
21 Lift and conveyor installation	-	-	-
22 Protective and communication installation	9,133	2	19
23 Special installations/services equipment	13,700	3	28
24 Builders' work	41,111	9	84
Totals	456,680	100	932

Category: Public service
Type: Police station
Floor area: 342m2
Total cost: £337,896

	Elemental cost £	% of cost	Cost m2 floor area £
1 Preliminaries	37,168	11	108
2 Substructure	23,652	7	69
3 Frame	16,894	5	49
4 Upper floors	6,760	2	20
5 Roof	23,652	7	69
6 Staircases	10,136	3	30
7 External walls	37,168	11	108
8 Windows and external doors	23,652	7	69
9 Partitions and internal walls	13,515	4	39
10 Internal doors	10,136	3	30
11 Wall finishes	10,136	3	30
12 Floor finishes	13,515	4	39
13 Ceiling finishes	6,760	2	20
14 Fittings and furnishings	6,760	2	20
15 Sanitary appliances/disposal installation	6,760	2	20
16 Hot and cold water services	6,760	2	20
17 Heating and air treatment installation	6,760	2	20
18 Ventilation installation	3,378	1	10
19 Gas services	3,378	1	10
20 Electric installation	27,031	8	79
21 Lift and conveyor installation	-	-	-
22 Protective and communication installation	10,131	3	30
23 Special installations/services equipment	6,760	2	20
24 Builders' work	27,034	8	79
Totals	337,896	100	988

Category: Public service
Type: Village hall
Floor area: 380m2
Total cost: £262,960

	Elemental cost £	% of cost	Cost m2 floor area £
1 Preliminaries	28,925	11	75
2 Substructure	26,296	10	69
3 Frame	-	-	-
4 Upper floors	-	-	-
5 Roof	23,666	9	62
6 Staircases	-	-	-
7 External walls	26,296	10	69
8 Windows and external doors	18,407	7	48
9 Partitions and internal walls	10,518	4	28
10 Internal doors	7,888	3	21
11 Wall finishes	7,888	3	21
12 Floor finishes	10,518	4	28
13 Ceiling finishes	5,259	2	14
14 Fittings and furnishings	13,148	5	35
15 Sanitary appliances/disposal installation	5,259	2	14
16 Hot and cold water services	2,629	1	7
17 Heating and air treatment installation	7,888	3	21
18 Ventilation installation	-	-	-
19 Gas services	-	-	-
20 Electric installation	31,562	12	83
21 Lift and conveyor installation	-	-	-
22 Protective and communication installation	7,888	3	21
23 Special installations/services equipment	2,629	1	7
24 Builders' work	26,296	10	69
Totals	262,960	100	692

Category: Industrial
Type: Factory, light industrial
Floor area: 13,820m2
Total cost: £5,500,360

	Elemental cost £	% of cost	Cost m2 floor area £
1 Preliminaries	385,025	7	28
2 Substructure	715,046	13	52
3 Frame	880,068	16	64
4 Upper floors	-	-	-
5 Roof	550,036	10	40
6 Staircases	-	-	-
7 External walls	550,036	10	40
8 Windows and external doors	220,014	4	15
9 Partitions and internal walls	165,010	3	12
10 Internal doors	55,003	1	4
11 Wall finishes	55,003	1	4
12 Floor finishes	110,007	2	8
13 Ceiling finishes	55,003	1	4
14 Fittings and furnishings	440,028	8	32
15 Sanitary appliances/disposal installation	55,003	1	4
16 Hot and cold water services	55,003	1	4
17 Heating and air treatment installation	220,014	4	15
18 Ventilation installation	55,003	1	4
19 Gas services	55,003	1	4
20 Electric installation	440,028	8	32
21 Lift and conveyor installation	-	-	-
22 Protective and communication installation	330,021	6	26
23 Special installations/services equipment	55,003	1	4
24 Builders' work	55,003	1	4
Totals	5,500,360	100	398

Category: Industrial
Type: Factory, heavy industrial
Floor area: 10,500m2
Total cost: £4,567,500

	Elemental cost £	% of cost	Cost m2 floor area £
1 Preliminaries	411,075	9	39
2 Substructure	456,750	10	44
3 Frame	548,100	12	52
4 Upper floors	182,700	4	17
5 Roof	411,075	9	39
6 Staircases	45,675	1	4
7 External walls	365,400	8	35
8 Windows and external doors	137,025	3	13
9 Partitions and internal walls	91,350	2	9
10 Internal doors	45,675	1	4
11 Wall finishes	91,350	2	9
12 Floor finishes	91,350	2	9
13 Ceiling finishes	45,675	1	4
14 Fittings and furnishings	45,675	1	4
15 Sanitary appliances/disposal installation	91,350	2	9
16 Hot and cold water services	91,350	2	9
17 Heating and air treatment installation	182,700	4	17
18 Ventilation installation	182,700	4	17
19 Gas services	91,350	2	9
20 Electric installation	456,750	10	44
21 Lift and conveyor installation	-	-	-
22 Protective and communication installation	91,350	2	9
23 Special installations/services equipment	91,350	2	9
24 Builders' work	319,725	7	30
Totals	4,567,500	100	435

Category: Industrial
Type: Livestock building
Floor area: 800m2
Total cost: £209,600

	Elemental cost £	% of cost	Cost m2 floor area £
1 Preliminaries	23,056	11	29
2 Substructure	14,672	7	18
3 Frame	14,672	7	18
4 Upper floors	-	-	-
5 Roof	16,768	8	21
6 Staircases	-	-	-
7 External walls	33,536	16	42
8 Windows and external doors	6,288	3	8
9 Partitions and internal walls	10,480	5	13
10 Internal doors	2,096	1	3
11 Wall finishes	-	-	-
12 Floor finishes	6,288	3	8
13 Ceiling finishes	-	-	-
14 Fittings and furnishings	-	-	-
15 Sanitary appliances/disposal installation	-	-	-
16 Hot and cold water services	4,192	2	5
17 Heating and air treatment installation	-	-	-
18 Ventilation installation	2,096	1	3
19 Gas services	-	-	-
20 Electric installation	33,536	16	42
21 Lift and conveyor installation	-	-	-
22 Protective and communication installation	4,192	2	5
23 Special installations/services equipment	14,672	7	18
24 Builders' work	23,056	11	29
Totals	209,600	100	262

Category: Industrial
Type: Laboratory
Floor area: 940m2
Total cost: £913,680

	Elemental cost £	% of cost	Cost m2 floor area £
1 Preliminaries	82,231	9	87
2 Substructure	54,820	6	58
3 Frame	63,958	7	68
4 Upper floors	63,958	7	68
5 Roof	73,094	8	78
6 Staircases	18,273	2	19
7 External walls	54,820	6	58
8 Windows and external doors	45,684	5	49
9 Partitions and internal walls	36,547	4	39
10 Internal doors	27,410	3	29
11 Wall finishes	18,273	2	19
12 Floor finishes	27,410	3	29
13 Ceiling finishes	27,410	3	29
14 Fittings and furnishings	36,547	4	39
15 Sanitary appliances/disposal installation	9,140	1	10
16 Hot and cold water services	9,140	1	10
17 Heating and air treatment installation	9,140	1	10
18 Ventilation installation	9,140	1	10
19 Gas services	9,140	1	10
20 Electric installation	100,504	11	107
21 Lift and conveyor installation	-	-	-
22 Protective and communication installation	18,273	2	19
23 Special installations/services equipment	27,410	3	29
24 Builders' work	91,368	10	98
Totals	913,680	100	972

Category: Industrial
Type: Nursery units
Floor area: 2,018m2
Total cost: £710,336

	Elemental cost £	% of cost	Cost m2 floor area £
1 Preliminaries	85,246	12	42
2 Substructure	71,033	10	35
3 Frame	-	-	-
4 Upper floors	-	-	-
5 Roof	113,653	16	56
6 Staircases	-	-	-
7 External walls	156,273	22	77
8 Windows and external doors	21,310	3	10
9 Partitions and internal walls	14,206	2	7
10 Internal doors	7,103	1	4
11 Wall finishes	7,103	1	4
12 Floor finishes	14,206	2	7
13 Ceiling finishes	7,103	1	4
14 Fittings and furnishings	-	-	-
15 Sanitary appliances/disposal installation	7,103	1	4
16 Hot and cold water services	7,103	1	4
17 Heating and air treatment installation	14,206	2	7
18 Ventilation installation	-	-	-
19 Gas services	7,103	1	4
20 Electric installation	78,136	11	38
21 Lift and conveyor installation	-	-	-
22 Protective and communication installation	14,206	2	7
23 Special installations/services equipment	-	-	-
24 Builders' work	85,246	12	42
Totals	710,336	100	352

Category: Industrial
Type: Warehouse shell
Floor area: 2,864m2
Total cost: £887,840

	Elemental cost £	% of cost	Cost m2 floor area £
1 Preliminaries	88,784	10	31
2 Substructure	124,297	14	44
3 Frame	142,054	16	51
4 Upper floors	8,878	1	3
5 Roof	115,419	13	40
6 Staircases	8,878	1	3
7 External walls	115,419	13	40
8 Windows and external doors	44,392	5	16
9 Partitions and internal walls	17,756	2	6
10 Internal doors	17,756	2	6
11 Wall finishes	8,878	1	3
12 Floor finishes	26,635	3	9
13 Ceiling finishes	8,878	1	3
14 Fittings and furnishings	8,878	1	3
15 Sanitary appliances/disposal installation	8,878	1	3
16 Hot and cold water services	8,878	1	3
17 Heating and air treatment installation	17,756	2	6
18 Ventilation installation	-	-	-
19 Gas services	-	-	-
20 Electric installation	62,157	7	22
21 Lift and conveyor installation	-	-	-
22 Protective and communication installation	8,878	1	3
23 Special installations/services equipment	8,878	1	3
24 Builders' work	35,513	4	12
Totals	887,840	100	310

Category: Industrial
Type: Warehouse, complete
Floor area: 2,120m2
Total cost: £905,240

	Elemental cost £	% of cost	Cost m2 floor area £
1 Preliminaries	81,471	9	38
2 Substructure	99,576	11	47
3 Frame	117,681	13	55
4 Upper floors	9,052	1	4
5 Roof	117,681	13	55
6 Staircases	9,052	1	4
7 External walls	117,681	13	56
8 Windows and external doors	36,219	4	17
9 Partitions and internal walls	18,104	2	9
10 Internal doors	18,104	2	9
11 Wall finishes	18,104	2	9
12 Floor finishes	27,157	3	13
13 Ceiling finishes	18,104	2	9
14 Fittings and furnishings	18,104	2	9
15 Sanitary appliances/disposal installation	9,052	1	4
16 Hot and cold water services	9,052	1	4
17 Heating and air treatment installation	27,157	3	13
18 Ventilation installation	-	-	-
19 Gas services	9,052	1	4
20 Electric installation	63,366	7	30
21 Lift and conveyor installation	-	-	-
22 Protective and communication installation	27,157	3	13
23 Special installations/services equipment	9,052	1	4
24 Builders' work	45,262	5	21
Totals	905,240	100	427

Category: Health and welfare
Type: Group surgery
Floor area: 395m2
Total cost: £288,745

	Elemental cost £	% of cost	Cost m2 floor area £
1 Preliminaries	34,649	12	88
2 Substructure	28,874	10	73
3 Frame	-	-	-
4 Upper floors	-	-	-
5 Roof	28,874	10	73
6 Staircases	-	-	-
7 External walls	34,649	12	88
8 Windows and external doors	17,324	6	43
9 Partitions and internal walls	17,324	6	43
10 Internal doors	8,662	3	22
11 Wall finishes	5,776	2	15
12 Floor finishes	8,662	3	22
13 Ceiling finishes	5,775	2	15
14 Fittings and furnishings	5,775	2	15
15 Sanitary appliances/disposal installation	5,775	2	15
16 Hot and cold water services	2,887	1	7
17 Heating and air treatment installation	2,887	1	7
18 Ventilation installation	2,887	1	7
19 Gas services	-	-	-
20 Electric installation	31,765	11	80
21 Lift and conveyor installation	-	-	-
22 Protective and communication installation	5,775	2	15
23 Special installations/services equipment	5,775	2	15
24 Builders' work	34,649	12	88
Totals	288,745	100	731

Category: Health and welfare
Type: Health centre
Floor area: 510m2
Total cost: £452,370

	Elemental cost £	% of cost	Cost m2 floor area £
1 Preliminaries	49,760	11	98
2 Substructure	45,237	10	89
3 Frame	-	-	-
4 Upper floors	-	-	-
5 Roof	49,760	11	98
6 Staircases	-	-	-
7 External walls	58,808	13	113
8 Windows and external doors	18,094	4	35
9 Partitions and internal walls	18,094	4	35
10 Internal doors	13,571	3	27
11 Wall finishes	9,047	2	18
12 Floor finishes	13,571	3	27
13 Ceiling finishes	9,047	2	18
14 Fittings and furnishings	18,094	4	35
15 Sanitary appliances/disposal installation	13,571	3	27
16 Hot and cold water services	4,523	1	9
17 Heating and air treatment installation	4,523	1	9
18 Ventilation installation	4,523	1	9
19 Gas services	-	-	-
20 Electric installation	54,293	12	106
21 Lift and conveyor installation	-	-	-
22 Protective and communication installation	9,047	2	18
23 Special installations/services equipment	9,047	2	18
24 Builders' work	49,760	11	98
Totals	452,370	100	887

Category: Health and welfare
Type: Old persons' nursing home
Floor area: 1,672m2
Total cost: £1,524,864

		Elemental cost £	% of cost	Cost m2 floor area £
1	Preliminaries	167,735	11	100
2	Substructure	121,989	8	73
3	Frame	-	-	-
4	Upper floors	-	-	-
5	Roof	198,241	13	119
6	Staircases	-	-	-
7	External walls	91,492	6	55
8	Windows and external doors	106,740	7	64
9	Partitions and internal walls	60,994	4	37
10	Internal doors	76,243	5	45
11	Wall finishes	60,994	4	37
12	Floor finishes	60,994	4	37
13	Ceiling finishes	30,497	2	18
14	Fittings and furnishings	91,492	6	55
15	Sanitary appliances/disposal installation	45,745	3	27
16	Hot and cold water services	45,745	3	27
17	Heating and air treatment installation	106,740	7	64
18	Ventilation installation	15,248	1	9
19	Gas services	15,248	1	9
20	Electric installation	91,492	6	55
21	Lift and conveyor installation	-	-	-
22	Protective and communication installation	45,745	3	27
23	Special installations/services equipment	45,745	3	27
24	Builders' work	45,745	3	27
	Totals	1,524,864	100	912

Category: Health and welfare
Type: Welfare centre
Floor area: 612m2
Total cost: £521,424

		Elemental cost £	% of cost	Cost m2 floor area £
1	Preliminaries	62,570	12	101
2	Substructure	67,795	13	111
3	Frame	-	-	-
4	Upper floors	-	-	-
5	Roof	57,356	11	94
6	Staircases	-	-	-
7	External walls	62,570	12	101
8	Windows and external doors	20,856	4	34
9	Partitions and internal walls	15,642	3	26
10	Internal doors	15,642	3	26
11	Wall finishes	15,642	3	26
12	Floor finishes	15,642	3	26
13	Ceiling finishes	10,428	2	17
14	Fittings and furnishings	26,071	5	41
15	Sanitary appliances/disposal installation	10,428	2	17
16	Hot and cold water services	5,214	1	9
17	Heating and air treatment installation	10,428	2	17
18	Ventilation installation	5,214	1	9
19	Gas services	-	-	-
20	Electric installation	57,356	11	94
21	Lift and conveyor installation	-	-	-
22	Protective and communication installation	10,428	2	17
23	Special installations/services equipment	5,214	1	9
24	Builders' work	46,928	9	77
	Totals	521,424	100	852

Category: Leisure
Type: Golf club house
Floor area: 675m2
Total cost: £465,750

	Elemental cost £	% of cost	Cost m2 floor area £
1 Preliminaries	46,575	10	69
2 Substructure	32,602	7	48
3 Frame	-	-	-
4 Upper floors	-	-	-
5 Roof	65,210	14	97
6 Staircases	-	-	-
7 External walls	51,232	11	75
8 Windows and external doors	18,630	4	28
9 Partitions and internal walls	13,972	3	21
10 Internal doors	13,972	3	21
11 Wall finishes	13,972	3	21
12 Floor finishes	9,315	2	14
13 Ceiling finishes	9,315	2	14
14 Fittings and furnishings	37,260	8	55
15 Sanitary appliances/disposal installation	9,315	2	14
16 Hot and cold water services	4,657	1	7
17 Heating and air treatment installation	4,657	1	7
18 Ventilation installation	-	-	-
19 Gas services	4,657	1	7
20 Electric installation	65,205	14	97
21 Lift and conveyor installation	-	-	-
22 Protective and communication installation	9,315	2	14
23 Special installations/services equipment	4,657	1	7
24 Builders' work	51,232	11	75
Totals	465,750	100	690

Category: Leisure
Type: Public house
Floor area: 780m2
Total cost: £577,200

	Elemental cost £	% of cost	Cost m2 floor area £
1 Preliminaries	51,948	9	84
2 Substructure	34,632	6	56
3 Frame	11,544	2	18
4 Upper floors	11,544	2	18
5 Roof	57,720	10	93
6 Staircases	17,316	3	28
7 External walls	34,632	6	56
8 Windows and external doors	23,088	4	37
9 Partitions and internal walls	17,316	3	28
10 Internal doors	23,088	4	37
11 Wall finishes	28,860	5	46
12 Floor finishes	28,860	5	46
13 Ceiling finishes	23,088	4	37
14 Fittings and furnishings	69,264	12	111
15 Sanitary appliances/disposal installation	17,316	3	28
16 Hot and cold water services	5,772	1	9
17 Heating and air treatment installation	23,088	4	37
18 Ventilation installation	5,772	1	9
19 Gas services	5,772	1	9
20 Electric installation	40,404	7	6
21 Lift and conveyor installation	-	-	-
22 Protective and communication installation	11,544	2	18
23 Special installations/services equipment	23,088	4	37
24 Builders' work	11,544	2	18
Totals	577,200	100	740

Category: Leisure
Type: Restaurant
Floor area: 340m2
Total cost: £307,700

		Elemental cost £	% of cost	Cost m2 floor area £
1	Preliminaries	27,693	9	82
2	Substructure	24,616	8	72
3	Frame	-	-	-
4	Upper floors	-	-	-
5	Roof	33,847	11	100
6	Staircases	-	-	-
7	External walls	40,001	13	118
8	Windows and external doors	18,462	6	54
9	Partitions and internal walls	21,539	7	63
10	Internal doors	18,462	6	54
11	Wall finishes	12,308	4	36
12	Floor finishes	15,385	5	45
13	Ceiling finishes	9,231	3	27
14	Fittings and furnishings	21,539	7	64
15	Sanitary appliances/disposal installation	3,077	1	9
16	Hot and cold water services	3,077	1	9
17	Heating and air treatment installation	3,077	1	9
18	Ventilation installation	3,077	1	9
19	Gas services	3,077	1	9
20	Electric installation	30,770	10	91
21	Lift and conveyor installation	-	-	-
22	Protective and communication installation	6,154	2	18
23	Special installations/services equipment	3,077	1	9
24	Builders' work	9,231	3	27
	Totals	307,700	100	905

Category: Leisure
Type: Sports hall
Floor area: 2,0162m
Total cost: £1,334,592

		Elemental cost £	% of cost	Cost m2 floor area £
1	Preliminaries	80,075	6	38
2	Substructure	133,459	10	66
3	Frame	160,152	12	81
4	Upper floors	-	-	-
5	Roof	146,805	11	73
6	Staircases	-	-	-
7	External walls	93,421	7	47
8	Windows and external doors	40,037	3	20
9	Partitions and internal walls	40,037	3	20
10	Internal doors	26,693	2	13
11	Wall finishes	26,693	2	13
12	Floor finishes	93,421	7	47
13	Ceiling finishes	13,346	1	7
14	Fittings and furnishings	106,767	8	53
15	Sanitary appliances/disposal installation	53,383	4	27
16	Hot and cold water services	26,693	2	13
17	Heating and air treatment installation	53,383	4	26
18	Ventilation installation	13,346	1	7
19	Gas services	-	-	-
20	Electric installation	106,767	8	53
21	Lift and conveyor installation	-	-	-
22	Protective and communication installation	26,693	2	13
23	Special installations/services equipment	13,346	1	7
24	Builders' work	80,075	6	38
	Totals	1,334,592	100	662

Category: Education
Type: Library
Floor area: 616m2
Total cost: £437,360

	Elemental cost £	% of cost	Cost m2 floor area £
1 Preliminaries	52,483	12	85
2 Substructure	39,362	9	64
3 Frame	-	-	-
4 Upper floors	-	-	-
5 Roof	52,483	12	85
6 Staircases	-	-	-
7 External walls	61,237	14	101
8 Windows and external doors	30,615	7	50
9 Partitions and internal walls	8,747	2	14
10 Internal doors	8,747	2	14
11 Wall finishes	13,120	3	21
12 Floor finishes	13,120	3	21
13 Ceiling finishes	8,746	2	14
14 Fittings and furnishings	39,362	9	64
15 Sanitary appliances/disposal installation	4,373	1	7
16 Hot and cold water services	4,373	1	7
17 Heating and air treatment installation	8,747	2	14
18 Ventilation installation	-	-	-
19 Gas services	-	-	-
20 Electric installation	34,988	8	57
21 Lift and conveyor installation	-	-	-
22 Protective and communication installation	8,747	2	14
23 Special installations/services equipment	4,373	1	7
24 Builders' work	43,736	10	71
Totals	437,360	100	710

Category: Education
Type: Primary school
Floor area: 2,108m2
Total cost: £1,675,860

	Elemental cost £	% of cost	Cost m2 floor area £
1 Preliminaries	150,827	9	71
2 Substructure	117,310	7	56
3 Frame	-	-	-
4 Upper floors	-	-	-
5 Roof	167,586	10	80
6 Staircases	-	-	-
7 External walls	184,344	11	86
8 Windows and external doors	201,111	12	94
9 Partitions and internal walls	100,551	6	48
10 Internal doors	67,034	4	32
11 Wall finishes	50,275	3	24
12 Floor finishes	67,034	4	32
13 Ceiling finishes	33,517	2	16
14 Fittings and furnishings	100,551	6	48
15 Sanitary appliances/disposal installation	67,034	4	32
16 Hot and cold water services	16,758	1	8
17 Heating and air treatment installation	67,034	4	32
18 Ventilation installation	-	-	-
19 Gas services	16,758	1	8
20 Electric installation	134,068	8	64
21 Lift and conveyor installation	-	-	-
22 Protective and communication installation	33,517	2	16
23 Special installations/services equipment	16,758	1	8
24 Builders' work	83,793	5	40
Totals	1,675,860	100	795

Category: Education
Type: Secondary school
Floor area: 2,450m2
Total cost: £1,817,900

	Elemental cost £	% of cost	Cost m2 floor area £
1 Preliminaries	163,611	9	67
2 Substructure	127,253	7	52
3 Frame	-	-	-
4 Upper floors	-	-	-
5 Roof	199,969	11	80
6 Staircases	-	-	-
7 External walls	181,790	10	73
8 Windows and external doors	163,611	9	67
9 Partitions and internal walls	90,895	5	37
10 Internal doors	72,716	4	30
11 Wall finishes	54,537	3	22
12 Floor finishes	72,716	4	30
13 Ceiling finishes	36,358	2	15
14 Fittings and furnishings	163,611	9	67
15 Sanitary appliances/disposal installation	36,358	2	15
16 Hot and cold water services	18,179	1	8
17 Heating and air treatment installation	54,537	3	22
18 Ventilation installation	-	-	-
19 Gas services	-	-	-
20 Electric installation	163,611	9	67
21 Lift and conveyor installation	-	-	-
22 Protective and communication installation	36,358	2	15
23 Special installations/services equipment	18,179	1	8
24 Builders' work	163,611	9	67
Totals	1,817,900	100	742

Category: Education
Type: Sixth form college
Floor area: 1,828m2
Total cost: £1,312,504

	Elemental cost £	% of cost	Cost m2 floor area £
1 Preliminaries	131,250	10	72
2 Substructure	105,000	8	57
3 Frame	-	-	-
4 Upper floors	-	-	-
5 Roof	118,125	9	66
6 Staircases	-	-	-
7 External walls	157,502	12	86
8 Windows and external doors	105,000	8	57
9 Partitions and internal walls	65,625	5	36
10 Internal doors	52,500	4	29
11 Wall finishes	39,375	3	22
12 Floor finishes	39,375	3	22
13 Ceiling finishes	26,250	2	14
14 Fittings and furnishings	157,502	12	86
15 Sanitary appliances/disposal installation	13,125	1	7
16 Hot and cold water services	13,125	1	7
17 Heating and air treatment installation	39,375	3	22
18 Ventilation installation	-	-	-
19 Gas services	-	-	-
20 Electric installation	105,000	8	57
21 Lift and conveyor installation	-	-	-
22 Protective and communication installation	26,250	2	14
23 Special installations/services equipment	13,125	1	7
24 Builders' work	105,000	8	57
Totals	1,312,504	100	718

Category: Education
Type: Special school
Floor area: 1,710m2
Total cost: £1,265,400

	Elemental cost £	% of cost	Cost m2 floor area £
1 Preliminaries	139,194	11	81
2 Substructure	113,886	9	67
3 Frame	-	-	-
4 Upper floors	-	-	-
5 Roof	126,540	10	74
6 Staircases	-	-	-
7 External walls	139,194	11	81
8 Windows and external doors	88,578	7	52
9 Partitions and internal walls	63,270	5	37
10 Internal doors	37,962	3	22
11 Wall finishes	37,962	3	22
12 Floor finishes	37,962	3	22
13 Ceiling finishes	25,308	2	15
14 Fittings and furnishings	177,156	14	104
15 Sanitary appliances/disposal installation	25,308	2	15
16 Hot and cold water services	12,654	1	7
17 Heating and air treatment installation	37,962	3	22
18 Ventilation installation	-	-	-
19 Gas services	-	-	-
20 Electric installation	101,232	8	59
21 Lift and conveyor installation	-	-	-
22 Protective and communication installation	25,308	2	15
23 Special installations/services equipment	25,308	2	15
24 Builders' work	50,616	4	30
Totals	**1,265,400**	**100**	**740**

Category: Education
Type: Teachers' training college
Floor area: 2,620m2
Total cost: £1,637,500

	Elemental cost £	% of cost	Cost m2 floor area £
1 Preliminaries	163,750	10	62
2 Substructure	131,000	8	50
3 Frame	98,250	6	37
4 Upper floors	65,500	4	25
5 Roof	163,750	10	62
6 Staircases	49,125	3	19
7 External walls	147,375	9	56
8 Windows and external doors	98,250	6	37
9 Partitions and internal walls	65,500	4	25
10 Internal doors	65,500	4	25
11 Wall finishes	49,125	3	19
12 Floor finishes	49,125	3	19
13 Ceiling finishes	32,750	2	13
14 Fittings and furnishings	114,625	7	44
15 Sanitary appliances/disposal installation	32,750	2	13
16 Hot and cold water services	16,375	1	6
17 Heating and air treatment installation	32,750	2	13
18 Ventilation installation	-	-	-
19 Gas services	-	-	-
20 Electric installation	131,000	8	50
21 Lift and conveyor installation	-	-	-
22 Protective and communication installation	32,750	2	13
23 Special installations/services equipment	16,375	1	6
24 Builders' work	81,875	5	31
Totals	1,637,500	100	625

Category: Residential
Type: Local Authority low rise flats
Floor area: 2,349m2
Total cost: £1,348,326

	Elemental cost £	% of cost	Cost m2 floor area £
1 Preliminaries	134,832	10	57
2 Substructure	134,832	10	57
3 Frame	-	-	-
4 Upper floors	53,933	4	23
5 Roof	148,319	11	64
6 Staircases	26,966	2	11
7 External walls	161,799	12	69
8 Windows and external doors	107,870	8	49
9 Partitions and internal walls	67,416	5	28
10 Internal doors	53,933	4	23
11 Wall finishes	40,449	3	17
12 Floor finishes	40,449	3	17
13 Ceiling finishes	67,416	5	28
14 Fittings and furnishings	40,449	3	17
15 Sanitary appliances/disposal installation	26,966	2	11
16 Hot and cold water services	13,483	1	6
17 Heating and air treatment installation	80,899	6	34
18 Ventilation installation	13,483	1	6
19 Gas services	-	-	-
20 Electric installation	67,416	5	28
21 Lift and conveyor installation	-	-	-
22 Protective and communication installation	26,966	2	11
23 Special installations/services equipment	26,966	2	11
24 Builders' work	13,483	1	6
Totals	1,348,326	100	574

Category: Residential
Type: Local Authority housing
Floor area: 3,988m2
Total cost: £2,133,580

	Elemental cost £	% of cost	Cost m2 floor area £
1 Preliminaries	192,022	9	48
2 Substructure	128,014	6	32
3 Frame	-	-	-
4 Upper floors	149,350	7	37
5 Roof	277,365	13	70
6 Staircases	64,007	3	16
7 External walls	213,358	10	54
8 Windows and external doors	128,014	6	32
9 Partitions and internal walls	85,343	4	21
10 Internal doors	64,007	3	16
11 Wall finishes	85,343	4	21
12 Floor finishes	42,671	2	11
13 Ceiling finishes	42,671	2	11
14 Fittings and furnishings	85,343	4	21
15 Sanitary appliances/disposal installation	42,670	2	11
16 Hot and cold water services	42,670	2	11
17 Heating and air treatment installation	64,007	3	16
18 Ventilation installation	-	-	-
19 Gas services	-	-	-
20 Electric installation	298,709	14	75
21 Lift and conveyor installation	-	-	-
22 Protective and communication installation	-	-	-
23 Special installations/services equipment	-	-	-
24 Builders' work	128,014	6	32
Totals	2,133,580	100	535

Category: Residential
Type: Private flats
Floor area: 1,260m2
Total cost: £749,700

	Elemental cost £	% of cost	Cost m2 floor area £
1 Preliminaries	44,982	6	36
2 Substructure	44,982	6	36
3 Frame	-	-	-
4 Upper floors	37,485	5	30
5 Roof	89,964	12	71
6 Staircases	22,491	3	18
7 External walls	82,467	11	65
8 Windows and external doors	44,982	6	36
9 Partitions and internal walls	22,491	3	18
10 Internal doors	29,988	4	23
11 Wall finishes	22,491	3	18
12 Floor finishes	22,491	3	18
13 Ceiling finishes	22,491	3	18
14 Fittings and furnishings	29,988	4	23
15 Sanitary appliances/disposal installation	7,497	1	6
16 Hot and cold water services	14,994	2	12
17 Heating and air treatment installation	22,491	3	18
18 Ventilation installation	-	-	-
19 Gas services	-	-	-
20 Electric installation	82,467	11	56
21 Lift and conveyor installation	29,988	4	23
22 Protective and communication installation	22,491	3	18
23 Special installations/services equipment	7,497	1	6
24 Builders' work	44,982	6	36
Totals	749,700	100	595

Category: Residential
Type: Luxury private flats
Floor area: 2,400m2
Total cost: £2,100,000

	Elemental cost £	% of cost	Cost m2 floor area £
1 Preliminaries	147,000	7	61
2 Substructure	126,000	6	53
3 Frame	-	-	-
4 Upper floors	147,000	7	61
5 Roof	189,000	9	79
6 Staircases	63,000	3	26
7 External walls	210,000	10	88
8 Windows and external doors	147,000	7	61
9 Partitions and internal walls	63,000	3	26
10 Internal doors	84,000	4	35
11 Wall finishes	63,000	3	26
12 Floor finishes	42,000	2	17
13 Ceiling finishes	42,000	2	17
14 Fittings and furnishings	126,000	6	53
15 Sanitary appliances/disposal installation	42,000	2	17
16 Hot and cold water services	42,000	2	17
17 Heating and air treatment installation	63,000	3	26
18 Ventilation installation	-	-	
19 Gas services	-	-	-
20 Electric installation	210,000	10	88
21 Lift and conveyor installation	105,000	5	44
22 Protective and communication installation	63,000	3	26
23 Special installations/services equipment	21,000	1	10
24 Builders' work	105,000	5	44
Totals	2,100,000	100	875

Category: Transport
Type: Multi-storey car park
Floor area: 9,600m2
Total cost: £2,246,400

	Elemental cost £	% of cost	Cost m2 floor area £
1 Preliminaries	247,104	11	26
2 Substructure	292,032	13	31
3 Frame	471,744	21	49
4 Upper floors	404,352	18	42
5 Roof	22,464	1	2
6 Staircases	112,320	5	13
7 External walls	89,856	4	9
8 Windows and external doors	22,464	1	2
9 Partitions and internal walls	22,464	1	2
10 Internal doors	22,464	1	2
11 Wall finishes	-	-	-
12 Floor finishes	-	-	-
13 Ceiling finishes	-	-	-
14 Fittings and furnishings	-	-	-
15 Sanitary appliances/disposal installation	-	-	-
16 Hot and cold water services	-	-	-
17 Heating and air treatment installation	-	-	-
18 Ventilation installation	44,928	2	5
19 Gas services	-	-	-
20 Electric installation	202,176	9	21
21 Lift and conveyor installation	157,248	7	16
22 Protective and communication installation	67,392	3	7
23 Special installations/services equipment	44,928	2	5
24 Builders' work	22,464	1	2
Totals	2,246,400	100	234

MECHANICAL AND ELECTRICAL ELEMENTAL COSTS

		Office block	Hospital	Hotel
Mechanical work		£	£	£
1	Heating	35	59	72
2	Ventilation	24	46	71
3	Fire protection	18	21	27
4	Ancillary services	7	11	9
	Total	84	137	179
Electrical work				
1	Lighting	35	59	72
2	Mains	24	46	71
3	Power	18	21	27
4	Lightning protection	3	3	3
5	Communications	3	14	12
6	Security	4	12	11
7	Ancillary services	5	6	6
	Total	68	126	102

		Leisure complex	Sports stadium	Hotel
Mechanical work		£	£	£
1	Heating	32	2	33
2	Ventilation	64	2	16
3	Fire protection	18	8	11
4	Ancillary services	8	3	10
	Total	122	15	70
Electrical work				
1	Lighting	38	27	24
2	Mains	22	14	19
3	Power	14	9	18
4	Lightning protection	2	3	2
5	Communications	6	8	4
6	Security	10	9	10
7	Ancillary services	8	8	7
	Total	100	78	84

4

Composite and principle rates

When the appraisal of a project moves beyond the square metre stage, it is likely that the client will commission a cost plan to be prepared. This entails taking off rough quantities and applying global rates to produce an approximate cost of the project. These rates are the result of combining various item descriptions and costs into what are called composite rates.

Not all items can be combined with others to provide composite rates and principle rates are also included if their value is significant to the cost plan. The following rates are presented under headings in the same order as the elements in Chapter 3.

BUILDING WORK

PRELIMINARIES

These will usually be assessed by referring to the needs of each particular project but a figure of 7.5 to 12.5% of the construction costs should be added depending upon the nature of the project.

SUBSTRUCTURES

Excavation   £

Excavate by machine to reduce levels and
dispose of excavated material

deposit on site	m3	3.00
spread and level on site	m3	5.00
remove from site	m3	16.00

£

Excavate by machine for basements and
dispose of excavated material

deposit on site	m3	5.00
spread and level on site	m3	7.00
remove from site	m3	18.00

Breaking out and disposal

Extra over excavation for breaking out

rock	m3	35.00
concrete	m3	28.00
reinforced concrete	m3	40.00
brickwork or blockwork	m3	20.00

Extra over excavation for disposing of contaminated material to licensed tip	m3	40.00

Filling

Filling to make up levels including
levelling and compacting

excavated material	m3	5.00
sand	m3	30.00
hardcore	m3	25.00
DOT Type 1	m3	35.00
DOT Type 2	m3	33.00

Foundation walling

Excavate trench by machine, dispose of
surplus excavated material off site,
earthwork support, concrete foundations
10 N/mm2 40mm aggregate (1:3:6),
common bricks (£150/1000) in cavity wall,
pitch polymer damp proof course,
concrete foundation size

£

450 × 150mm
wall height 750mm	m	96.00
wall height 1000mm	m	108.00
wall height 1250mm	m	115.00

450 × 225mm
wall height 750mm	m	102.00
wall height 1000mm	m	114.00
wall height 1250mm	m	121.00

600 × 225mm
wall height 750mm	m	108.00
wall height 1000mm	m	120.00
wall height 1250mm	m	130.00

Excavate trench by machine, dispose of
surplus excavated material off site,
earthwork support, concrete foundations
10 N/mm2 40mm aggregate (1:3:6),
engineering bricks (£200/1000) in cavity wall
and damp proof course for foundation size

450 × 150mm
wall height 750mm	m	102.00
wall height 1000mm	m	114.00
wall height 1250mm	m	120.00

450 × 225mm
wall height 750mm	m	108.00
wall height 1000mm	m	120.00
wall height 1250mm	m	128.00

600 × 225mm
wall height 750mm	m	114.00
wall height 1000mm	m	130.00
wall height 1250mm	m	140.00

Bases, pile caps and ground beams £

Reinforced concrete (11.5N/mm2 40mm
aggregate) in bases including excavation,
earthwork support, formwork and
reinforcement, size

600 × 600 × 300mm	nr	44.00
750 × 750 × 450mm	nr	60.00
1000 × 1000 × 450mm	nr	80.00
1200 × 1200 × 600mm	nr	140.00

Reinforced concrete (21N/mm2 20mm
aggregate) in pile caps including
excavation, earthwork support,
formwork and reinforcement, size

750 × 750 × 600mm	nr	200.00
900 × 900 × 1000mm	nr	475.00
1000 × 1000 × 1000mm	nr	1100.00
2000 × 2000 × 1000mm	nr	1500.00

Reinforced concrete (21N/mm2 20mm
aggregate) in ground beams
including excavation, earthwork
support, formwork and reinforcement,
size

450 × 450mm	m	60.00
750 × 500mm	m	110.00
900 × 500mm	m	135.00
1000 × 1000mm	m	280.00

Strip foundations and slabs   **£**

Strip foundations including trench
excavation, disposal, earthwork
support, concrete foundation (11.5N/m2
40mm aggregate), brick cavity wall with
three courses of facing bricks, wall height

in commons (£150/1000)		
600mm	m	70.00
900mm	m	86.00
1200mm	m	102.00
1500mm	m	118.00
in facings (£250/1000)		
600mm	m	80 00
900mm	m	100.00
1200mm	m	120.00
1500mm	m	140.00
in engineering bricks (£230/1000)		
600mm	m	76.00
900mm	m	94.00
1200mm	m	112.00
1500mm	m	130.00

Ground slab including excavation,
disposal, hardcore bed blinded with
sand, 1200 gauge polythene damp proof
membrane, concrete (21Nmm2 20mm
aggregate)

one layer of reinforcement A252		
150mm	m2	34.00
200mm	m2	36.00
250mm	m2	38.00
300mm	m2	40.00

£

two layers of reinforcement A252

150mm	m2	37.00
200mm	m2	39.00
250mm	m2	41.00
300mm	m2	43.00

FRAME

Reinforced concrete (30N/mm2
20mm aggregate) including
reinforcement and formwork

columns, size

225 × 225mm	m	38.00
300 × 300mm	m	50.00
300 × 450mm	m	70.00
450 × 450mm	m	100.00
450 × 600mm	m	135.00
450 × 900mm	m	150.00

Reinforced concrete (30N/mm2
20mm aggregate) including
reinforcement and formwork

beams, size

225 × 225mm	m	48.00
300 × 450mm	m	90.00
300 × 600mm	m	110.00
450 × 450mm	m	100.00
450 × 600mm	m	135.00
450 × 900mm	m	160.00
600 × 900mm	m	210.00
600 × 1200mm	m	260.00

£

Fabricated galvanised steelwork
BS4360 grade 40 erected on site
including connections

columns	t	1900.00
universal beams	t	1650.00
roof bracing	t	2000.00
lattice beams	t	1750.00

UPPER FLOORS

Softwood flooring including 25mm
thick tongued and grooved boarding
on joists size

125 × 50mm	m2	28.00
150 × 50mm	m2	30.00
175 × 50mm	m2	32.00
200 × 50mm	m2	34.00
225 × 50mm	m2	36.00
250 × 50mm	m2	38.00

Softwood flooring including 18mm
thick chipboard flooring butt jointed
on joists size

125 × 50mm	m2	24.00
150 × 50mm	m2	26.00
175 × 50mm	m2	28.00
200 × 50mm	m2	30.00
225 × 50mm	m2	32.00
250 × 50mm	m2	34.00

Reinforced concrete suspended slabs
including formwork

150mm thick	m2	65.00
200mm thick	m2	75.00

£

Precast concrete suspended slabs

150mm thick	m2	80.00
200mm thick	m2	95.00

ROOFS

The following items represent
the costs of roofs measured on
plan. The costs for roof coverings
are given separately.

Flat roofs

Reinforced concrete suspended slabs
including formwork

150mm thick	m2	65.00
200mm thick	m2	75.00

Precast concrete suspended slabs

150mm thick	m2	80.00
200mm thick	m2	95.00

Softwood flat roofing consisting of
herringbone strutting, wall plates,
joists and woodwool slabs

woodwool slabs 50mm thick		
joists 150 × 50mm	m2	32.00
joists 175 × 50mm	m2	34.00
joists 200 × 50mm	m2	36.00
joists 250 × 50mm	m2	38.00
woodwool slabs 75mm thick		
joists 150 × 50mm	m2	38.00
joists 175 × 50mm	m2	40.00
joists 200 × 50mm	m2	42.00
joists 250 × 50mm	m2	44.00

£

woodwool slabs 100mm thick		
joists 150 × 50mm	m2	48.00
joists 175 × 50mm	m2	50.00
joists 200 × 50mm	m2	52.00
joists 250 × 50mm	m2	54.00

Flat roof decking

Reinforced woodwool slab decking
50mm thick covered with

two layer bituminous roofing	m2	35.00
three layer bituminous roofing	m2	40.00
two layer asphalt roofing	m2	55.00

Reinforced woodwool slab decking
75mm thick covered with

two layer bituminous roofing	m2	40.00
three layer bituminous roofing	m2	45.00
two layer asphalt roofing	m2	60.00

Reinforced woodwool slab decking
100mm thick covered with

two layer bituminous roofing	m2	50.00
three layer bituminous roofing	m2	55.00
two layer asphalt roofing	m2	70.00

Pitched roofs

Softwood trusses at 600mm centres

span 5m		
pitch 22.5 degrees	m2	24.00
pitch 30 degrees	m2	26.00
pitch 40 degrees	m2	28.00
pitch 45 degrees	m2	30.00

		£
span 8m		
pitch 22.5 degrees	m2	22.00
pitch 30 degrees	m2	24.00
pitch 40 degrees	m2	26.00
pitch 45 degrees	m2	28.00
span 10m		
pitch 22.5 degrees	m2	20.00
pitch 30 degrees	m2	22.00
pitch 40 degrees	m2	24.00
pitch 45 degrees	m2	26.00

Underlays

Roofing felt

unreinforced	m2	1.50
reinforced	m2	2.00

Building paper	m2	1.50

External quality plywood

12mm thick	m2	16.00
15mm thick	m2	19.00
18mm thick	m2	22.00

Chipboard

12mm thick	m2	5.00
15mm thick	m2	6.00
12mm thick	m2	8.00
22mm thick	m2	10.00

Insulation quilt

100mm thick	m2	6.00
150mm thick	m2	7.00
200mm thick	m2	8.00

Roof coverings £

The following costs include for underfelt, roof battens and work to eaves, verges and ridges The rates are based on sloping areas.

Welsh blue slates size

610 × 305mm	m2	60.00
510 × 255mm	m2	55.00
405 × 205mm	m2	65.00

Westmorland green slates size

610 × 305mm	m2	130.00
405 × 255mm	m2	120.00
355 × 205mm	m2	135.00

Reconstructed stone slates size

457 × 305mm	m2	45.00
457 × 457mm	m2	40.00

Asbestos-free slates

400 × 200mm	m2	55.00
500 × 250mm	m2	50.00
600 × 300mm	m2	45.00

Concrete interlocking tiles

420 × 330mm	m2	20.00
380 × 230mm	m2	24.00
325 × 330mm	m2	22.00
420 × 220mm	m2	21.00
430 × 380mm	m2	18.00

£

Clay pantiles

340 × 240mm	m2	34.00
470 × 280mm	m2	30.00

Eaves, fascias and verges

Painted softwood fascia 250mm
high and 16mm thick softwood
soffit

200mm wide	m	22.00
400mm wide	m	24.00

Painted softwood fascia 250mm
high and 12mm thick plywood
soffit

200mm wide	m	20.00
400mm wide	m	22.00

Painted softwood barge board
250mm high to sloping roof with
painted softwood soffit

150mm wide	m	18.00
250mm wide	m	22.00

Rainwater goods

The following costs include
for all fittings.

Rainwater pipes

cast iron		
75mm	m	28.00
100mm	m	34.00

		£
PVC-U		
68m	m	12.00
112mm	m	16.00
aluminium		
63mm	m	20.00
75mm	m	22.00
100mm	m	32.00

Rainwater gutters

cast iron		
100mm	m	22.00
125mm	m	14.00
PVC-U		
75mm	m	12.00
112mm	m	14.00
aluminium		
100mm	m	24.00
112mm	m	26.00
125mm	m	28.00

Sheet coverings

Lead sheeting

1.80mm thick	m2	60.00
2.24mm thick	m2	70.00
2.65mm thick	m2	76.00
3.15mm thick	m2	86.00
3.55mm thick	m2	90.00

Aluminium sheeting

0.90mm thick	m2	50.00

£

Copper sheeting

0.56mm thick	m2	60.00
0.61mm thick	m2	65.00

Zinc sheeting

0.80mm thick	m2	55.00

Bituminous built up roofing

two layer	m2	14.00
three layer	m2	18.00

Two coat asphalt

13mm thick	m2	14.00
20mm thick	m2	18.00
30mm thick	m2	24.00

Roof cladding

Asbestos-free corrugated single
skin cladding

grey	m2	22.00
coloured	m2	24.00

Asbestos-free corrugated double
skin cladding

grey	m2	56.00
coloured	m2	58.00

Lightweight galvanised steel
sheeting m2 24.00

Aluminium profiled cladding with pre-
painted finish m2 34.00

STAIRCASES

£

Reinforced concrete construction,
3250mm rise, granolithic finish
including mild steel painted
balustrade and handrails

Straight flight, width

900mm	nr	2850.00
1200mm	nr	3250.00

Dog-leg flight, width

900mm	nr	3300.00
1200mm	nr	3850.00

Reinforced concrete construction,
3250mm rise, terrazzo finish
including mild steel painted
balustrade and handrails

Straight flight, width

900mm	nr	3000.00
1200mm	nr	3450.00

Dog-leg flight, width

900mm	nr	3500.00
1200mm	nr	4000.00

Softwood construction, 2600mm rise

straight flight 900mm wide, no balustrade	nr	900.00
two flights with quarter landing, softwood balustrade	nr	1150.00

£

two flights with half landing, hardwood handrail	nr	1350.00

Mild steel construction, 3000mm rise (balustrades and handrails excluded)

straight flight 900mm wide	nr	950.00
two flights with quarter landing	nr	1200.00
two flights with half landing	nr	1350.00

Spiral staircase in steel including powder coated mild steel balustrades and handrail, 3500mm overall rise

1500mm diameter	nr	2400.00
2000mm diameter	nr	2800.00

EXTERNAL WALLS

Solid brickwork/blockwork

Common bricks £120 per 1000

half brick thick	m2	30.00
one brick thick	m2	50.00
one and a half brick thick	m2	75.00

Common bricks £150 per 1000

half brick thick	m2	34.00
one brick thick	m2	55.00
one and a half brick thick	m2	80.00

		£
Facing bricks £300 per 1000		
half brick thick	m2	40.00
one brick thick	m2	70.00
Facing bricks £450 per 1000		
half brick thick	m2	80.00
Engineering bricks £200 per 1000		
half brick thick	m2	35.00
one brick thick	m2	60.00
one and a half brick thick	m2	85.00
Engineering bricks £250 per 1000		
half brick thick	m2	38.00
one brick thick	m2	65.00
one and a half brick thick	m2	95.00
Precast concrete blocks		
solid		
75mm thick	m2	15.00
100mm thick	m2	18.00
150mm thick	m2	23.00
hollow		
100mm thick	m2	18.00
140mm thick	m2	20.00
215mm thick	m2	26.00
Lightweight concrete blocks in wall		
100mm thick	m2	18.00
140mm thick	m2	22.00
190mm thick	m2	32.00

£

Dense concrete blocks in wall

100mm thick	m2	16.00
140mm thick	m2	32.00

Reconstructed stone blocks in wall

100mm thick	m2	44.00

Concrete walls

Concrete (21N/m 20mm aggregate) in wall
including reinforcement

sawn formwork
100mm thick	m2	70.00
150mm thick	m2	75.00
200mm thick	m2	80.00
250mm thick	m2	85.00
300mm thick	m2	90.00

wrought formwork
100mm thick	m2	75.00
150mm thick	m2	80.00
200mm thick	m2	85.00
250mm thick	m2	90.00
300mm thick	m2	95.00

Composite walls

Cavity wall formed with
lightweight block inner skin
100mm thick, stainless steel
wall ties, 50mm thick insulation
and outer skin of

common brickwork 102mm thick (£120 per 1000)	m2	56.00

		£
common brickwork 102mm thick (£150 per 1000)	m2	60.00
facing brickwork 102mm thick (£300 per 1000)	m2	66.00
facing brickwork 102mm thick (£450 per 1000)	m2	75.00
engineering brickwork 102mm thick (£200 per 1000)	m2	66.00
engineering brickwork 102mm thick (£250 per 1000)	m2	70.00
reconstructed stone block 100mm thick	m2	80.00
in situ concrete wall 100mm thick	m2	96.00
in situ concrete wall 150mm thick	m2	100.00
in situ concrete wall 200mm thick	m2	106.00
in situ concrete wall 250mm thick	m2	114.00
in situ concrete wall 300mm thick	m2	125.00

Cavity wall insulation

25mm thick board	m2	10.00
50mm thick board	m2	12.00

Wall cladding

Precast concrete panels	m2	136.00
Precast concrete panels with e×posed aggregate finish	m2	180.00
Precast concrete panels with reconstituted stone facing	m2	250.00

		£
Precast concrete panels with Portland stone natural facing	m2	440.00
GRP panels including associated fixings, insulation and flashings		
single skin panels	m2	180.00
double skin panels	m2	230.00
Non-asbestos coloured profiled cladding including insulation	m2	30.00

Metal cladding systems

PVF2 coated profiled galvanised steel sheeting, 80mm thick insulation and coated inner lining	m2	50.00
Insulated PVF2 coated silver finished metal sandwich panel system	m2	90.00

Curtain walling

Single glazed powder coated aluminium framed curtain walling system	m2	300.00
Double glazed powder coated aluminium framed curtain walling system	m2	450.00
Extra over curtain walling systems for		
dummy panels	m2	60.00
Antisun finish to external skin	m2	90.00

External finishes

		£
Two coat render with painted finish	m2	14.00
Two coat masonry paint	m2	6.00
Clay tile hanging including battens and felt	m2	38.00
Boarding in Western Red Cedar	m2	34.00

WINDOWS AND EXTERNAL DOORS

Windows

Softwood painted windows

single glazed	m2	190.00
double glazed	m2	240.00

Hardwood painted windows

single glazed	m2	290.00
double glazed	m2	330.00

Steel painted windows

single glazed	m2	200.00
double glazed	m2	250.00

Aluminium painted windows

single glazed	m2	250.00
double glazed	m2	390.00

PVC-U windows

double glazed	m2	380.00

£

External doors (including frames and ironmongery)

Softwood framed, ledged and braced door	nr	200.00
Softwood flush door	nr	240.00
Softwood panelled door	nr	330.00
Hardwood flush door	nr	490.00
Hardwood four panelled door	nr	580.00

Steel roller shutters and grilles

Manual	m2	240.00
Electric	m2	340.00
Grilles	m2	140.00

PARTITIONS AND INTERNAL WALLS

Common bricks £150 per 1000

half brick thick	m2	34.00
one brick thick	m2	55.00

Precast concrete blocks

solid

75mm thick	m2	15.00
100mm thick	m2	18.00
150mm thick	m2	23.00

hollow

100mm thick	m2	18.00
140mm thick	m2	20.00
215mm thick	m2	26.00

£

Lightweight concrete blocks

100mm thick	m2	18.00
140mm thick	m2	22.00
190mm thick	m2	32.00

Dense concrete blocks

100mm thick	m2	16.00
140mm thick	m2	32.00

Concrete (21N/m 20mm aggregate) in wall
including reinforcement

sawn formwork

100mm thick	m2	70.00
150mm thick	m2	75.00

wrought formwork

100mm thick	m2	75.00
150mm thick	m2	80.00

Stud partition faced with

plasterboard 9.5mm thick

one side	m2	35.00
two sides	m2	52.00

plasterboard 12.5mm thick

one side	m2	60.00
two sides	m2	82.00

One skim coat of plaster

one side	m2	5.00
two sides	m2	10.00

One mist and two coats of emulsion

one side	m2	3.00
two sides	m2	6.00

£

Metal stud partition, faced both sides
with plasterboard, one hour fire
resistance

170mm thick	m2	50.00
200mm thick	m2	60.00

Demountable steel framed
partitions

vinyl faced	m2	160.00
glazed	m2	150.00

Softwood framed glazed screens	m2	190.00

Hardwood framed glazed screens	m2	240.00

INTERNAL DOORS (including frames and ironmongery)

Softwood flush door 40mm thick size
762 × 1981mm

plywood faced	nr	150.00
sapele faced	nr	160.00

Softwood flush door 40mm thick size
826 × 2040mm

plywood faced	nr	160.00
sapele faced	nr	170.00

Softwood flush door 40mm thick,
half hour fire resistant, size
762 × 1981mm

plywood faced	nr	180.00
sapele faced	nr	190.00

£

Softwood flush door 40mm thick,
half hour fire resistant, size
826 × 2040mm

plywood faced	nr	190.00
sapele faced	nr	200.00

Purpose made softwood door
44mm thick size 762 × 1981mm

four panelled	nr	220.00

Purpose made softwood door
44mm thick size 826 × 2040mm

four panelled	nr	245.00

Purpose made mahogany door
with one panel open for
glass, size

762 × 1981 × 50mm	nr	240.00
838 × 1981 × 50mm	nr	250.00
762 × 1981 × 63mm	nr	270.00
838 × 1981 × 63mm	nr	280.00

Purpose made mahogany door
with two panels open for
glass, size

762 × 1981 × 50mm	nr	280.00
838 × 1981 × 50mm	nr	300.00
762 × 1981 × 63mm	nr	320.00
838 × 1981 × 63mm	nr	330.00

WALL FINISHES   £

In situ finishings

One coat plaster board finish 5mm thick to plasterboard	m2	4.00
Two coat plaster 10mm thick to concrete	m2	7.00
One coat plaster to brick or block walls		
13mm thick	m2	6.00
19mm thick	m2	8.00
Cement rendering in two coats to brick or block walls		
12mm thick	m2	7.00
18mm thick	m2	8.00
Plasterboard 9.5mm thick and skim coat	m2	12.00
Plasterboard 12.5mm thick and skim coat	m2	15.00
One mist and two coats of emulsion	m2	3.00
One coat undercoat and two coats gloss	m2	10.00
Lining paper	m2	3.00
Vinyl paper		
(PC £4.00/roll)	m2	4.00
(PC £6.00/roll)	m2	5.00
(PC £8.00/roll)	m2	6.00

Board linings   £

Dry plasterboard lining to walls
for direct decoration with
emulsion paint finish

9.5mm wallboard	m2	11.00
12.5mm wallboard	m2	12.00

Sheet linings on softwood battens
plugged and screwed to walls

plywood 4mm thick	m2	20.00
plywood 6mm thick	m2	22.00
plywood 9mm thick	m2	24.00
hardboard 3.2mm thick	m2	14.00
hardboard 4.8mm thick	m2	15.00
hardboard 6.0mm thick	m2	16.00
wallboard 9.5mm thick	m2	16.00
wallboard 12.5mm thick	m2	17.00
chipboard 12mm thick	m2	13.00
chipboard 15mm thick	m2	14.00
chipboard 18mm thick	m2	15.00
chipboard 25mm thick	m2	19.00
softwood boarding 13mm thick	m2	24.00
softwood boarding 25mm thick	m2	30.00

Tiling

Glazed ceramic wall tiling fixed with
adhesive and grouted

108 × 108 × 4mm	m2	36.00
152 × 152 × 5.5mm	m2	34.00
200 × 200 × 7mm	m2	32.00

FLOOR FINISHES   £

Timber flooring

Softwood flooring, butt jointed

19mm thick	m2	18.00
22mm thick	m2	20.00
25mm thick	m2	22.00

Softwood flooring, tongued
and grooved

19mm thick	m2	20.00
22mm thick	m2	22.00
25mm thick	m2	24.00

Hardwood strip flooring, tongued
and grooved, 25mm thick

utile mahogany	m2	50.00
maple	m2	58.00
iroko	m2	54.00

Chipboard flooring, butt
jointed

18mm thick	m2	10.00
22mm thick	m2	12.00

Chipboard flooring, tongued
and grooved

18mm thick	m2	12.00
22mm thick	m2	14.00

Plywood flooring, butt jointed

15mm thick	m2	22.00
18mm thick	m2	26.00

		£

Plywood flooring, tongued and grooved

15mm thick	m2	24.00
18mm thick	m2	28.00

Screeds/in situ finishings

Cement and sand (1:3) screed

25mm thick	m2	7.00
38mm thick	m2	9.00
50mm thick	m2	12.00
75mm thick	m2	14.00

Latex screed

3mm thick	m2	6.00
5mm thick	m2	8.00

Granolithic screed

25mm thick	m2	16.00
32mm thick	m2	18.00
38mm thick	m2	22.00
50mm thick	m2	26.00

Mastic flooring

black

20mm thick	m2	18.00
25mm thick	m2	22.00

red

20mm thick	m2	22.00
25mm thick	m2	26.00

Epoxy floor finish, 5mm thick	m2	26.00

Tile flooring   £

Quarry tiles

    red

150 × 150 × 12.5mm	m2	32.00
200 × 200 × 25mm	m2	42.00

    brown

150 × 150 × 12.5mm	m2	42.00
200 × 200 × 25mm	m2	52.00

Ceramic tiles

    red

100 × 100 × 9mm	m2	32.00
150 × 150 × 12mm	m2	38.00
200 × 200 × 12mm	m2	50.00

    black

100 × 100 × 9mm	m2	36.00
150 × 150 × 12mm	m2	42.00
200 × 200 × 12mm	m2	54.00

Flexible flooring

Rubber floor tiles, 3mm thick

    plain finish

black	m2	44.00
brown	m2	46.00

    studded finish

black	m2	50.00
brown	m2	52.00

Linoleum sheeting

2.5mm thick	m2	14.00
3.2mm thick	m2	18.00

		£
Linoleum tiling 3.2mm thick	m2	20.00
Thermoplastic tiling, 2mm thick	m2	9.00
Vinyl sheeting		
2mm thick	m2	12.00
2.5mm thick	m2	14.00
3mm thick	m2	15.00
Vinyl tiling, 2mm thick	m2	10.00
Cork tiling		
3.2mm thick	m2	15.00
4.8mm thick	m2	18.00
6.3mm thick	m2	22.00
8.0mm thick	m2	30.00
Woodblock flooring herringbone pattern, sanded and wax polished		
maple	m2	60.00
oak	m2	60.00
merbau	m2	54.00
iroko	m2	58.00
Wood strip flooring herringbone pattern, sanded and wax polished		
maple	m2	56.00
oak	m2	56.00
merbau	m2	50.00
iroko	m2	52.00
Underlay to carpets		
rubber	m2	4.00
felt	m2	3.00

£

Fitted carpets

			£
contract quality			
medium duty		m2	12.00
heavy duty		m2	16.00

Carpet tiles

domestic	m2	16.00
medium duty	m2	26.00
heavy duty	m2	38.00

Skirtings

Softwood painted skirtings	m2	9.00
Hardwood stained skirtings	m2	13.00

Vinyl coved skirtings

75mm high	m2	4.00
100mm high	m2	5.00

CEILING FINISHES

In situ finishings

Skim coat of plaster to ceilings	m2	4.00
One coat plaster board finish 5mm thick to plasterboard	m2	5.00
One coat plaster board finish 8mm thick to concrete ceilings	m2	7.00
One mist and two coats of emulsion	m2	4.00

		£
One coat Artex sealer and one coat Artex finish		
plastered ceilings	m2	4.00
plasterboard ceilings	m2	4.00
Lining paper	m2	4.00
Vinyl paper		
(PC £4.00/roll)	m2	5.00
(PC £6.00/roll)	m2	6.00
(PC £8.00/roll)	m2	7.00

Board finishes

Plasterboard 9.5mm thick and skim coat	m2	13.00
Plasterboard 12.5mm thick and skim coat	m2	16.00

Suspended ceiling systems

Gyproc M/F suspended ceiling system with 12.7mm thick wallboard	m2	24.00
Armstrong suspended ceiling system with acoustic tiles		
300 × 300mm	m2	22.00
600 × 600mm	m2	20.00
Egg crate suspended ceiling	m2	76.00
Coffered feature ceiling	m2	120.00

FITTINGS AND FURNISHINGS   £

The range of sizes and quality in fittings
means that it is not possible to provide
reliable costs for this section.

SANITARY FITTINGS/DISPOSAL INSTALLATIONS

Sanitary fittings (complete with water supply, taps and waste pipework)

White

lavatory basin	nr	180.00
WC	nr	160.00
urinal bowl	nr	175.00
shower cubicle	nr	420.00
sink	nr	300.00
bath, steel enamelled	nr	340.00
bath, acrylic	nr	300.00

Coloured

lavatory basin	nr	210.00
WC	nr	190.00
shower cubicle	nr	450.00
bath, acrylic	nr	340.00

Pipework (including all fittings)

Waste pipes

copper		
35mm	m	16.00
42mm	m	20.00
polypropylene		
35mm	m	7.00
42mm	m	8.00

		£
PVC-U		
19mm	m	6.00
32mm	m	8.00
38mm	m	9.00

Overflow pipes

copper		
22mm	m	8.00
28mm	m	10.00
polypropylene		
· 35mm	m	7.00
42mm	m	8.00
PVC-U		
19mm	m	6.00

Soil pipes

cast iron		
175mm	m	40.00
100mm	m	45.00
PVC-U		
100mm	m	22.00

HOT AND COLD WATER SERVICES

Pipework (including all fittings)

Copper pipes, capillary fittings

15mm	m	7.00
22mm	m	8.00
28mm	m	10.00
35mm	m	16.00
42mm	m	20.00
54mm	m	24.00

		£
Copper pipes, compression fittings		
15mm	m	9.00
22mm	m	10.00
28mm	m	12.00
35mm	m	19.00
42mm	m	24.00
54mm	m	28.00

Cisterns and cylinders (including all connections and overflows)

Galvanised iron cold water cisterns, capacity

36 litres	nr	80.00
86 litres	nr	100.00
191 litres	nr	140.00
327 litres	nr	180.00
491 litres	nr	300.00

Polyethylene cold water cisterns, capacity

68 litres	nr	100.00
114 litres	nr	120.00
182 litres	nr	130.00
227 litres	nr	140.00

GRP cold water cisterns, capacity

68 litres	nr	210.00
114 litres	nr	240.00
182 litres	nr	270.00
227 litres	nr	310.00

£

Copper cylinders, indirect pattern,
capacity

114 litres	nr	190.00
117 litres	nr	195.00
140 litres	nr	220.00
162 litres	nr	280.00

Copper cylinders, direct pattern,
capacity

116 litres	nr	180.00
120 litres	nr	230.00
144 litres	nr	280.00
166 litres	nr	330.00

HEATING AND AIR TREATMENT INSTALLATIONS

VENTILATION INSTALLATIONS

GAS SERVICES

ELECTRICAL INSTALLATION

LIFT AND CONVEYOR INSTALLATIONS

PROTECTION AND COMMUNICATION INSTALLATIONS

SPECIAL INSTALLATIONS/ SERVICES EQUIPMENT

See M&E Work, page 167 et seq.

BUILDERS WORK IN CONNECTION
WITH SPECIALIST SERVICES

Holes for pipes up to 55mm
diameter through   £

half brick wall	nr	5.00
one brick wall	nr	7.00
one and half brick wall	nr	10.00
blockwork 100mm thick	nr	4.00
blockwork 140mm thick	nr	6.00
blockwork 190mm thick	nr	8.00

Holes for pipes 55 to 110mm
diameter through

half brick wall	nr	6.00
one brick wall	nr	7.00
one and half brick wall	nr	11.00
blockwork 100mm thick	nr	5.00
blockwork 140mm thick	nr	7.00
blockwork 190mm thick	nr	9.00

Chases in walls for pipes
up to 55mm diameter

brickwork	m	4.00
blockwork	m	3.00

Chases in walls for pipes
55 to 110mm diameter

brickwork	m	7.00
blockwork	m	6.00

EXTERNAL WORKS

Excavation   £

Excavate topsoil 150mm thick
and deposit on site in spoil heaps

by machine	m2	2.00
by hand	m2	3.00

Excavate topsoil 150mm thick
and spread and level on site

by machine	m2	5.00
by hand	m2	12.00

Excavate to reduce levels and
deposit on site in spoil heaps

depth not exceeding 0.25m

by machine	m3	3.00
by hand	m3	24.00

depth not exceeding 1m

by machine	m3	4.00
by hand	m3	26.00

depth not exceeding 2m

by machine	m3	5.00
by hand	m3	28.00

depth not exceeding 4m

by machine	m3	6.00
by hand	m3	29.00

Excavate to reduce levels and
spread and level on site

depth not exceeding 0.25m

by machine	m3	4.00
by hand	m3	26.00

		£
depth not exceeding 1m		
by machine	m3	5.00
by hand	m3	28.00
depth not exceeding 2m		
by machine	m3	6.00
by hand	m3	30.00
depth not exceeding 4m		
by machine	m3	7.00
by hand	m3	33.00

Excavate to reduce levels and
remove to tip off site

depth not exceeding 0.25m		
by machine	m3	15.00
by hand	m3	40.00
depth not exceeding 1m		
by machine	m3	16.00
by hand	m3	42.00
depth not exceeding 2m		
by machine	m3	17.00
by hand	m3	44.00
depth not exceeding 4m		
by machine	m3	18.00
by hand	m3	45.00

Breaking up

Excavate in soft rock

by machine	m3	40.00
by hand	m3	55.00

		£
Excavate in hard rock		
by machine	m3	55.00
by hand	m3	65.00
Excavate in concrete		
by machine	m3	50.00
by hand	m3	60.00
Excavate in masonry		
by machine	m3	20.00
by hand	m3	25.00
Excavate in pavings 150mm thick		
macadam		
by machine	m2	3.00
by hand	m2	5.00
concrete		
by machine	m2	5.00
by hand	m2	7.00
macadam		
by machine	m2	8.00
by hand	m2	7.00

Filling

Imported filling material
deposited on site in layers not
exceeding 250mm thick, compacting
with vibrating roller

excavated material		
by machine	m3	15.00
by hand	m3	5.00

			£
sand			
	by machine	m3	32.00
	by hand	m3	37.00
hardcore			
	by machine	m3	24.00
	by hand	m3	32.00
granular fill Type 1			
	by machine	m3	32.00
	by hand	m3	38.00
granular fill Type 2			
	by machine	m3	30.00
	by hand	m3	36.00

Surface treatments

Level and compact

excavation bottom	m2	1.50
filling	m2	2.00

Trim

sloping excavated surfaces	m2	1.50
sloping rock	m2	8.00

Beds to receive pavings

Hardcore in bed

75mm thick		
by machine	m2	3.00
by hand	m2	3.50

100mm thick		
by machine	m2	3.50
by hand	m2	4.00

		£
150mm thick		
by machine	m2	4.00
by hand	m2	4.50

Quarry waste in bed

75mm thick		
by machine	m2	3.50
by hand	m2	4.00
100mm thick		
by machine	m2	4.00
by hand	m2	4.50
150mm thick		
by machine	m2	4.50
by hand	m2	5.00

Granular fill Type 1 in bed
by machine

100mm thick	m2	6.00
150mm thick	m2	7.00
200mm thick	m2	9.00
250mm thick	m2	11.00
300mm thick	m2	14.00

Pavings (laid on prepared bed)

In situ concrete with trowelled
finish

100mm thick	m2	8.00
125mm thick	m2	10.00
150mm thick	ı 2	12.00

£

Precast concrete paving flags
50mm thick

600 × 450mm			
	natural	m2	14.00
	coloured	m2	16.00
600 × 600mm			
	natural	m2	13.00
	coloured	m2	15.00
600 × 750mm			
	natural	m2	12.00
	coloured	m2	14.00
600 × 900mm			
	natural	m2	11.00
	coloured	m2	13.00

Concrete block paviours size
200 × 100 × 60mm thick, laid

straight bond			
	natural	m2	22.00
	coloured	m2	24.00
herringbone pattern			
	natural	m2	26.00
	coloured	m2	28.00

Concrete block paviours size
200 × 100 × 80mm thick, laid

straight bond			
	natural	m2	24.00
	coloured	m2	26.00
herringbone pattern			
	natural	m2	28.00
	coloured	m2	30.00

		£
Reconstructed stone paving flags		
size 450 × 450 × 40mm	m2	40.00
size 600 × 450 × 40mm	m2	38.00
Brick paviours size 215 × 102.5 × 65mm (£200 per 1000)		
laid flat		
straight bond	m2	24.00
herringbone pattern	m2	28.00
laid on edge		
straight bond	m2	32.00
herringbone pattern	m2	36.00
Brick paviours size 215 × 102.5 × 65mm (£400 per 1000)		
laid flat		
straight bond	m2	32.00
herringbone pattern	m2	36.00
laid on edge		
straight bond	m2	40.00
herringbone pattern	m2	44.00
Granite setts size 200 × 100 × 100mm	m2	60.00
Cobble paving average size 75mm	m2	75.00
York stone paving		
600 × 600 × 50mm thick	m2	90.00
600 × 600 × 50mm thick	m2	95.00

£

Fencing (including excavating for post holes, intermediate and end posts and concrete bases)

Chestnut fencing to BS1722 Part 4

1.00m high	m2	9.00
1.25m high	m2	10.00
1.50m high	m2	11.00
1.80m high	m2	12.00

Chainlink fencing to BS1722
Part 1 with galvanised wire
mesh, height

galvanised mild steel posts

0.90m	m2	15.00
1.20m	m2	18.00
1.80m	m2	22.00

concrete posts

0.90m	m2	14.00
1.20m	m2	17.00
1.80m	m2	20.00

Chainlink fencing to BS1722
Part 1 with plastic coated wire
mesh, height

galvanised mild steel posts

0.90m	m2	17.00
1.20m	m2	20.00
1.80m	m2	24.00

concrete posts

0.90m	m2	16.00
1.20m	m2	19.00
1.80m	m2	22.00

£

Close boarded fencing to BS1772 Part 5

on timber posts		
1.00m high	m	24.00
1.20m high	m	28.00
1.50m high	m	32.00
on concrete posts		
1.00m high	m	30.00
1.20m high	m	34.00
1.50m high	m	44.00

Panel fencing to BS1722 Part 11

on timber posts		
0.90m high	m	16.00
1.20m high	m	18.00
1.50m high	m	20.00
1.80m high	m	22.00
on concrete posts		
0.90m high	m	18.00
1.20m high	m	20.00
1.50m high	m	22.00
1.80m high	m	24.00

Palisade fencing to BS1722 Part 6

on timber posts		
0.90m high	m	18.00
1.20m high	m	20.00
1.50m high	m	22.00
1.80m high	m	24.00
on concrete posts		
0.90m high	m	22.00
1.20m high	m	24.00
1.50m high	m	26.00
1.80m high	m	28.00

£

Post and rail fencing to BS1722 Part 7

three rail morticed, 1.10m high	m	18.00
three rail nailed, 1.10m high	m	14.00
four rail morticed		
1.10m high	m	20.00
1.30m high	m	22.00
four rail nailed		
1.10m high	m	16.00
1.30m high	m	18.00

Drainage

Excavate trench by machine,
lay 100mm vitrified clay push fit
flexible pipe with

granular filling to		
bed and haunching		
depth 0.50m	m	27.00
depth 0.75m	m	29.00
depth 1.00m	m	32.00
depth 1.25m	m	37.00
depth 1.50m	m	42.00
depth 1.75m	m	47.00
depth 2.00m	m	52.00
depth 2.25m	m	57.00
depth 2.50m	m	62.00
depth 2.75m	m	67.00
depth 3.00m	m	72.00
granular filling to		
bed and surround		
depth 0.50m	m	31.00
depth 0.75m	m	33.00
depth 1.00m	m	36.00

£

depth 1.25m	m	41.00
depth 1.50m	m	46.00
depth 1.75m	m	51.00
depth 2.00m	m	56.00
depth 2.25m	m	61.00
depth 2.50m	m	66.00
depth 2.75m	m	71.00
depth 3.00m	m	76.00

concrete to bed and
haunching

depth 0.50m	m	33.00
depth 0.75m	m	35.00
depth 1.00m	m	38.00
depth 1.25m	m	43.00
depth 1.50m	m	48.00
depth 1.75m	m	53.00
depth 2.00m	m	58.00
depth 2.25m	m	63.00
depth 2.50m	m	68.00
depth 2.75m	m	73.00
depth 3.00m	m	78.00

concrete to bed and
surround

depth 0.50m	m	41.00
depth 0.75m	m	43.00
depth 1.00m	m	46.00
depth 1.25m	m	51.00
depth 1.50m	m	56.00
depth 1.75m	m	61.00
depth 2.00m	m	66.00
depth 2.25m	m	71.00
depth 2.50m	m	76.00
depth 2.75m	m	81.00
depth 3.00m	m	86.00

£

Excavate trench by machine,
lay 150mm vitrified clay push fit
flexible pipe with

granular filling to			
bed and haunching			
depth 0.50m		m	31.00
depth 0.75m		m	33.00
depth 1.00m		m	36.00
depth 1.25m		m	41.00
depth 1.50m		m	46.00
depth 1.75m		m	51.00
depth 2.00m		m	56.00
depth 2.25m		m	61.00
depth 2.50m		m	66.00
depth 2.75m		m	71.00
depth 3.00m		m	76.00
granular filling to			
bed and surround			
depth 0.50m		m	38.00
depth 0.75m		m	40.00
depth 1.00m		m	43.00
depth 1.25m		m	48.00
depth 1.50m		m	53.00
depth 1.75m		m	58.00
depth 2.00m		m	63.00
depth 2.25m		m	68.00
depth 2.50m		m	73.00
depth 2.75m		m	78.00
depth 3.00m		m	83.00
concrete to bed and			
haunching			
depth 0.50m		m	41.00
depth 0.75m		m	43.00
depth 1.00m		m	46.00
depth 1.25m		m	51.00
depth 1.50m		m	56.00

£

depth 2.00m	m	61.00
depth 2.25m	m	66.00
depth 2.50m	m	71.00
depth 2.75m	m	76.00
depth 3.00m	m	81.00

concrete to bed and
surround

depth 0.50m	m	49.00
depth 0.75m	m	51.00
depth 1.00m	m	54.00
depth 1.25m	m	59.00
depth 1.50m	m	64.00
depth 1.75m	m	69.00
depth 2.00m	m	74.00
depth 2.25m	m	79.00
depth 2.50m	m	84.00
depth 2.75m	m	89.00
depth 3.00m	m	94.00

Excavate trench by hand, lay
100mm vitrified clay push fit
flexible pipe with

granular filling to
bed and haunching

depth 0.50m	m	32.00
depth 0.75m	m	42.00
depth 1.00m	m	52.00
depth 1.25m	m	62.00
depth 1.50m	m	72.00
depth 1.75m	m	82.00
depth 2.00m	m	92.00
depth 2.25m	m	102.00
depth 2.50m	m	112.00
depth 2.75m	m	122.00
depth 3.00m	m	132.00

		£
granular filling to bed and surround		
depth 0.50m	m	36.00
depth 0.75m	m	46.00
depth 1.00m	m	56.00
depth 1.25m	m	66.00
depth 1.50m	m	76.00
depth 1.75m	m	86.00
depth 2.00m	m	96.00
depth 2.25m	m	106.00
depth 2.50m	m	116.00
depth 2.75m	m	126.00
depth 3.00m	m	136.00
concrete to bed and haunching		
depth 0.50m	m	38.00
depth 0.75m	m	48.00
depth 1.00m	m	58.00
depth 1.25m	m	68.00
depth 1.50m	m	78.00
depth 1.75m	m	88.00
depth 2.00m	m	98.00
depth 2.25m	m	108.00
depth 2.50m	m	118.00
depth 2.75m	m	128.00
depth 3.00m	m	138.00
concrete to bed and surround		
depth 0.50m	m	46.00
depth 0.75m	m	56.00
depth 1.00m	m	66.00
depth 1.25m	m	76.00
depth 1.50m	m	86.00
depth 1.75m	m	96.00
depth 2.00m	m	106.00
depth 2.25m	m	116.00
depth 2.50m	m	126.00
depth 2.75m	m	136.00
depth 3.00m	m	146.00

£

Manholes

Brick manholes including excavation,
concrete base, engineering brick walls,
main channel and bends,benching and
cast iron inspection cover, internal size

600 × 450mm		
depth, 500mm	nr	260.00
depth, 750mm	nr	330.00
depth, 1000mm	nr	400.00
depth, 1250mm	nr	460.00
depth, 1500mm	nr	520.00
750 × 450mm		
depth, 500mm	nr	300.00
depth, 750mm	nr	360.00
depth, 1000mm	nr	420.00
depth, 1250mm	nr	480.00
depth, 1500mm	nr	540.00
900 × 600mm		
depth, 500mm	nr	360.00
depth, 750mm	nr	440.00
depth, 1000mm	nr	520.00
depth, 1250mm	nr	600.00
depth, 1500mm	nr	680.00
depth, 1750mm	nr	760.00

Brick manholes including excavation,
concrete base, engineering brick walls,
main channel and bends, benching,
concrete reducing slab and cast iron
inspection cover, internal size

900 × 900mm		
depth, 750mm	nr	420.00
depth, 1000mm	nr	500.00

£

depth, 1250mm	nr	580.00
depth, 1500mm	nr	660.00
depth, 2000mm	nr	820.00
depth, 2500mm	nr	980.00
depth, 3000mm	nr	1140.00
depth, 3500mm	nr	1300.00
depth, 4000mm	nr	1460.00

1200 × 750mm

depth, 1000mm	nr	700.00
depth, 1250mm	nr	800.00
depth, 1500mm	nr	900.00
depth, 2000mm	nr	1100.00
depth, 2500mm	nr	1300.00
depth, 3000mm	nr	1500.00
depth, 3500mm	nr	1700.00
depth, 4000mm	nr	1900.00

900 × 1500mm

depth, 1000mm	nr	740.00
depth, 1250mm	nr	860.00
depth, 1500mm	nr	980.00
depth, 2000mm	nr	1220.00
depth, 2500mm	nr	1460.00
depth, 3000mm	nr	1700.00
depth, 3500mm	nr	1940.00
depth, 4000mm	nr	2180.00

1200 × 1800mm

depth, 1000mm	nr	760.00
depth, 1250mm	nr	890.00
depth, 1500mm	nr	1020.00
depth, 2000mm	nr	1280.00
depth, 2500mm	nr	1740.00
depth, 3000mm	nr	1920.00
depth, 3500mm	nr	2120.00
depth, 4000mm	nr	2380.00

LANDSCAPING

Site clearance   £

Demolish existing buildings
including digging up foundations

brick		
small	m3	10.00
medium	m3	8.00
large	m3	4.00
steel framed with cladded walls		
small	m3	4.00
medium	m3	3.00
large	m3	2.00
timber framed with cladded walls		
small	m3	3.00
medium	m3	2.00
large	m3	1.50

Temporary chestnut paling 1.50m high	m	10.00
Clear away scrub vegetation, shrub and hedges	m2	0.50

Cut down trees, grub up roots

trees size less than 600mm girth	nr	20.00
trees size 600 to 900mm girth	nr	60.00
trees size 900 to 1200mm girth	nr	100.00
trees size 1200 to 1500mm girth	nr	200.00
trees size 1500 to 1800mm girth	nr	260.00
trees size 1800 to 2100mm girth	nr	300.00
trees size 2100 to 2400mm girth	nr	450.00
trees size 2400 to 2700mm girth	nr	600.00
trees size 2700 to 3000mm girth	nr	900.00

£

Backfill tree hole with excavated
material

trees size less than 600mm girth	nr	2.00
trees size 600 to 900mm girth	nr	3.00
trees size 900 to 1200mm girth	nr	4.50
trees size 1200 to 1500mm girth	nr	6.00
trees size 1500 to 1800mm girth	nr	8.00
trees size 1800 to 2100mm girth	nr	9.00
trees size 2100 to 2400mm girth	nr	10.00
trees size 2400 to 2700mm girth	nr	15.00
trees size 2700 to 3000mm girth	nr	18.00

Backfill tree hole with hardcore

trees size less than 600mm girth	nr	6.00
trees size 600 to 900mm girth	nr	10.00
trees size 900 to 1200mm girth	nr	15.00
trees size 1200 to 1500mm girth	nr	20.00
trees size 1500 to 1800mm girth	nr	25.00
trees size 1800 to 2100mm girth	nr	30.00
trees size 2100 to 2400mm girth	nr	36.00
trees size 2400 to 2700mm girth	nr	40.00
trees size 2700 to 3000mm girth	nr	60.00

Backfill tree hole with sand

trees size less than 600mm girth	nr	10.00
trees size 600 to 900mm girth	nr	15.00
trees size 900 to 1200mm girth	nr	22.00
trees size 1200 to 1500mm girth	nr	30.00
trees size 1500 to 1800mm girth	nr	40.00
trees size 1800 to 2100mm girth	nr	45.00
trees size 2100 to 2400mm girth	nr	55.00
trees size 2400 to 2700mm girth	nr	65.00
trees size 2700 to 3000mm girth	nr	75.00

Excavation and filling   £

Excavate topsoil 150mm thick
and deposit on site in spoil heaps

by machine	m2	2.00
by hand	m2	3.00

Excavate topsoil 150mm thick
and spread and level on site

by machine	m3	20.00
by hand	m3	25.00

Imported filling material
deposited on site in layers not
exceeding 250mm thick, compacting
with vibrating roller

excavated material

by machine	m2	15.00
by hand	m2	25.00

sand

by machine	m2	31.00
by hand	m2	37.00

hardcore

by machine	m2	40.00
by hand	m2	30.00

granular fill Type 1

by machine	m2	32.00
by hand	m2	38.00

granular fill Type 2

by machine	m2	30.00
by hand	m2	36.00

Surface treatments   £

Break up existing ground
with plough or rotovator,
depth

100mm	m2	1.30
200mm	m2	1.50
300mm	m2	1.70
400mm	m2	1.90

Roll cultivated ground
with self-propelled roller   m2   0.60

Level and compact

excavation bottom	m2	1.50
filling	m2	2.00

Trim surfaces of

sloping excavated surfaces	m2	1.50
sloping rock	m2	8.00

Soil stabilisation

Trim existing sloping after
removal of topsoil, Tensar
matting pegged to sloping
grounds, covered with topsoil,
seeded and fertilised   m2   6.00

Polypropylene sheeting laid on
prepared ground   m2   2.40

Tensar erosion mats pegged to
ground with pins to prepared
ground   m2   4.50

		£
Greenfix bio-degradable erosion control mats fixed with pins to prepared ground		
seeded	m2	3.50
unseeded	m2	4.00
Grassblock precast concrete paving, laid on sand bed, filled in with topsoil and seeded		
83mm thick	m2	16.00
103mm thick	m2	20.00
125mm thick	m2	24.00
Grasscrete polystyrene formers, concrete infilling, topsoil and seeding		
100mm thick	m2	24.00
150mm thick	m2	28.00

Retaining walls

Excavate for and place in position wire mesh gabion cages filled with broken stone and rock, size		
zinc mesh		
2 × 1 × 0.5m	nr	80.00
2 × 1 × 1m	nr	100.00
PVC coated wire mesh		
2 × 1 × 0.5m	nr	90.00
2 × 1 × 1m	nr	110.00

£

Excavate for and place
in position galvanised wire
mesh gabion mattresses
filled with broken stone
and rock, size

6 × 2 × 0.25m	m2	24.00
6 × 2 × 0 30m	m2	30.00

Excavate trench by machine, dispose of
surplus excavated material off site,
earthwork support, concrete foundations
10 N/mm2 40mm aggregate (1:3:6),
engineering bricks (£200/1000) in
retaining wall

one brick thick wall

wall height 1500mm	m	200.00
wall height 2000mm	m	260.00
wall height 2500mm	m	320.00

one and a half brick thick wall

wall height 1500mm	m	260.00
wall height 2000mm	m	320.00
wall height 2500mm	m	380.00

Roads

Excavate for and lay road base,
tarmacadam sub-base and wearing
course, precast concrete kerbs both
sides, road width

5m	m	180.00
6m	m	220.00
7m	m	250.00

Paths

£

Excavate for hardcore base,
blinded with sand, covered in
bark chippings laid between
softwood edging boards

width, 1.5m	m	20.00
width, 2.0m	m	24.00
width, 2.5m	m	28.00
width, 3.0m	m	32.00

Excavate for hardcore base,
blinded with sand, lay precast
concrete blocks as fire path
infilled with topsoil and seeded

width, 3.0m	m	150.00
width, 4.0m	m	190.00
width, 5.0m	m	240.00

Precast concrete flags size
600 × 600 × 50mm laid
separately as stepping stones
including excavation and
sand bed

	nr	6.00

Car parks

Excavate and lay sub-base
to receive

tarmacadam 120mm thick		
general area	m2	40.00
parking bay	nr	480.00
concrete blocks 80mm thick		
general area	m2	35.00
parking bay	nr	420.00

£

cellular precast concrete
paving, filled in with topsoil
and seeded

83mm thick			
	general area	m2	16.00
	parking bay	nr	192.00
103mm thick			
	general area	m2	20.00
	parking bay	nr	240.00
125mm thick			
	general area	m2	24.00
	parking bay	nr	288.00

Sports grounds

Trim and grade prepared
groundand apply weedkiller,
fertiliser grass seed including
harrowing, rolling and one cut

general areas	m2	0.25
football pitch	nr	2390.00
rugby pitch	nr	1590.00
hockey pitch	nr	1150.00

Soiling, seeding and turfing

Imported topsoil filling spread
and levelled, average thickness

by machine			
	100mm	m2	2.00
	150mm	m2	3.00
	200mm	m2	3.50
	250mm	m2	4.00

		£
by hand		
100mm	m2	3.00
150mm	m2	4.00
200mm	m2	5.00
250mm	m2	6.50

Topsoil filling from spoil
heaps on site, spread and
levelled, average thickness

by machine		
100mm	m2	0.60
150mm	m2	0.80
200mm	m2	1.00
250mm	m2	1.20

by hand		
100mm	m2	1.60
150mm	m2	2.30
200mm	m2	3.00
250mm	m2	3.60

Plough and harrow topsoil to fine tilth, remove stones apply weedkiller	m2	0.20

Apply pesticide and weedkiller to
prepared topsoil, sow grass seed
(£2.50/kg), harrow, roll and one cut

by machine		
35g/m2	m2	0.15
50g/m2	m2	0.20
100g/m2	m2	0.35
150g/m2	m2	0.50
200g/m2	m2	0.70
300g/m2	m2	0.85
400g/m2	m2	1.10
500g/m2	m2	1.35

		£
by hand		
35g/m2	m2	0.20
50g/m2	m2	0.25
100g/m2	m2	0.40
150g/m2	m2	0.60
200g/m2	m2	0.80
300g/m2	m2	0.95
400g/m2	m2	1.20
500g/m2	m2	1.45

Imported turf size 4000 × 2000
× 19mm on prepared bed

general sports use	m2	3.00
special sports use	m2	3.50
domestic	m2	3.00

Treat turf with wooden
paddle beater   m2   0.20

Treat turf with light roller   m2   1.00

First cut to turf 20mm high
with man-operated power
driven cylinder mower
including boxing cuttings   m2   0.30

Planting

Excavate for and plant
transplants or seedlings,
backfill and water, plant cost

£0.50	nr	1.00
£1.00	nr	1.50
£1.50	nr	2.00
£2.00	nr	2.50
£2.50	nr	3.00
£3.00	nr	3.50

£

Excavate tree pit, fork bottom, plant
tree, backfill, water, surround with peat
(1 stake and 2 ties per tree), tree cost

£5.00	nr	10.00
£10.00	nr	18.00
£15.00	nr	25.00
£20.00	nr	40.00
£50.00	nr	100.00
£75.00	nr	150.00
£100.00	nr	240.00

Form planting hole in cultivated
area, plant shrub, backfill, water,
surround with peat, shrub cost

£5.00	nr	8.00
£10.00	nr	15.00

Tree stake 100mm diameter and
2 ties

length, 2m	nr	8.00
length, 2.5m	nr	9.00
length, 3m	nr	10.00

Galvanised wire tree guards,
300mm diameter

height, 1m	nr	15.00
height, 1.5m	nr	18.00
height, 2m	nr	22.00

Cast iron tree grille, size

two part

1000 × 1000mm	nr	175.00
1200 × 1200mm	nr	225.00
1200mm diameter	nr	200.00

		£
four part		
1000 × 1000mm	nr	225.00
1200 × 1200mm	nr	275.00
1200mm diameter	nr	250.00

Field drains

Excavate vee-sided ditch, bottom width

300mm, depth		
750mm	m	7.20
1000mm	m	9.60

500mm, depth		
750mm	m	12.00
1000mm	m	16.00
1250mm	m	20.00
1500mm	m	24.00

Excavate trench by machine and lay
agricultural clay pipes, backfill with
gravel rejects

75mm diameter		
depth, 400mm	m	9.40
depth, 500mm	m	10.80
depth, 600mm	m	12.60
depth, 700mm	m	14.20
depth, 800mm	m	15.80
depth, 900mm	m	17.40
depth, 1000mm	m	19.00

100mm diameter		
depth, 400mm	m	10.40
depth, 500mm	m	11.80
depth, 600mm	m	13.60
depth, 700mm	m	15.20
depth, 800mm	m	16.80
depth, 900mm	m	18.40
depth, 1000mm	m	20.00

			£
150mm diameter			
depth, 400mm	m		13.40
depth, 500mm	m		14.80
depth, 600mm	m		16.60
depth, 700mm	m		18.20
depth, 800mm	m		19.80
depth, 900mm	m		21.40
depth, 1000mm	m		23.00

Excavate trench by hand
and lay agricultural clay pipes

75mm diameter			
depth, 400mm	m		11.80
depth, 500mm	m		13.50
depth, 600mm	m		15.20
depth, 700mm	m		16.90
depth, 800mm	m		18.60
depth, 900mm	m		20.40
depth, 1000mm	m		22.00
100mm diameter			
depth, 400mm	m		12.80
depth, 500mm	m		14.50
depth, 600mm	m		16.20
depth, 700mm	m		17.90
depth, 800mm	m		19.60
depth, 900mm	m		21.40
depth, 1000mm	m		23.00
150mm diameter			
depth, 400mm	m		15.80
depth, 500mm	m		17.50
depth, 600mm	m		19.20
depth, 700mm	m		20.90
depth, 800mm	m		22.60
depth, 900mm	m		24.40
depth, 1000mm	m		26.00

CIVIL ENGINEERING WORK

Item rates in civil engineering are different from those in building because the percentage value of the General Items can be as high as 50% of the total construction cost. In building work the Preliminaries would usually be around 10%.

This is mainly due to the freedom of civil estimators to prepare bids to reflect their view of the timing and sequence of how the project would be constructed. This freedom is not usually available to building estimators.

So when preparing first stage estimates it is important to uplift the rates to allow for General Items or price them separately from the information presented below in Class A: General Items

CLASS A: GENERAL ITEMS

The items listed below are based upon the needs of a civil project whose contract value is approximately £5m with an 18 month (78 weeks) contract period.

Contractual requirements	£	£
Performance bond		
1% for construction period × 1.5 years × £5m	75,000	
0.75% for maintenance period × 1 year × £5m	<u>37,500</u>	102,500
Insurance of the Works		
1.5% × £5m		75,000
Insurance of construction plant		
included in hire charges		nil

	£	£

Insurance against damage to persons
and property

    included in head office overheads   nil

Specified requirements

Offices for Engineer's staff

	£	£
erect	500	
maintain and operate (100 weeks × £100)		10,000
remove	<u>500</u>	11,000

Laboratory for Engineer's staff

	£	£
erect	250	
maintain and operate (50 weeks × £75)	3,750	
remove	<u>250</u>	4,250

Cabins for Engineer's staff

	£	£
erect	250	
maintain and operate (100 weeks × £75)	7,500	
remove	<u>250</u>	8,000

Services for the Engineer

	£	£
1750cc car (100 weeks × £100)	10,000	

	£	£
Landrover (78 weeks × £100)	7,800	
telephone, installation	200	
telephone, maintain and operate (100 weeks × £60)	<u>6,000</u>	24,000

Equipment for the Engineer

	£	£
office equipment (100 weeks × £75) comprising	7,500	

    1 word processor
    2 desks
    2 tables
    1 conference table
    8 chairs
    2 filing cabinets
    sundries

	£	£
laboratory equipment (50 weeks × £75	3,750	
surveying equipment (78 weeks × £50)	<u>3,900</u>	15,150

Attendance upon the Engineer

	£	£
driver (50 weeks × £250)	12,500	
chainman (78 weeks × £200)	15,600	
laboratory assistant, part time 30 × £300)	<u>9,000</u>	37,100

		£
Testing of materials (included)		nil
Testing of works (included)		nil

	£	£

Temporary works

traffic signals (30 weeks @ £25)	750	
cleaning roads (40 weeks @ £300)	12,000	
progress photographs	1,000	
temporary lighting (40 weeks @ £40)	1,600	
temporary water supply, connection	1,000	
temporary water supply, pipework (200m × £6)	1,200	
temporary water supply (2,000,000 litres × 40p per 1000)	800	
temporary water supply, remove hardstanding (500m2 × £8)	4,000	
hardstanding, remove	<u>500</u>	22,850

Method Related Charges

Offices for contractor

erect	350
maintain and operate (100 weeks × £100)	10,000
remove	300

Cabins for contractor

erect	250
maintain and operate (78 weeks × £40)	3,120
remove	200

	£	£
Stores for contractor		
erect	250	
maintain and operate		
(100 weeks × £40)	4,000	
remove	200	
Canteens and messroom for contractor		
erect	300	
maintain and operate		
(78 weeks × £70)	5,460	
remove	250	
Electricity		
install	1,000	
maintain (100 weeks		
× £40)	4,000	
Supervision, administration		
agent 90 weeks × £500	45,000	
assistant agents (2)		
156 weeks × £400	62,400	
inspectors (2) 156 weeks		
× £350	54,600	
setting out engineer 30 weeks		
× £400	12,000	
quantity surveyor 110 weeks		
× £450	49,500	
section foremen (4)		
312 weeks × £350	109,200	
imekeeper/wages clerk		
78 weeks × £250	19,500	
storekeeper 78 weeks		
× £250	19,500	
watchman 100 weeks × £250	25,000	

	£	£
tea boy 78 weeks × £200	15,600	
Supervision, offloading and cleaning gang		
labourers (2) 140 weeks × £250	<u>35,000</u>	<u>476,980</u>
Total		<u>701,830</u>

CLASS B: GROUND INVESTIGATION

Trial pits

Trial pits size 1 × 2m, not
in rock, maximum depth

not exceeding 1m	nr	20.00
1-2m	nr	30.00
2-3m	nr	40.00
3-5m	nr	50.00

Trial pits size 1 × 2m, partly
in rock, maximum depth

not exceeding 1m	nr	30.00
1-2m	nr	40.00
2-3m	nr	50.00
3-5m	nr	60.00

Trial pits size 1 × 2m, in rock,
maximum depth

not exceeding 1m	nr	80.00
1-2m	nr	160.00
2-3m	nr	240.00
3-5m	nr	400.00

£

Light percussion boreholes,
150mm diameter

depth, not exceeding 5m	m	25.00
depth, 5-10m	m	30.00
depth, 10-20m	m	40.00
depth, 20-30m	m	60.00
depth, 30-40m	m	100.00

Rotary drilled boreholes, 150mm
diameter without core recovery

depth, not exceeding 5m	m	25.00
depth, 5-10m	m	30.00
depth, 10-20m	m	35.00
depth, 20-30m	m	40.00
depth, 30-40m	m	50.00

Rotary drilled boreholes,
75mm diameter, with core
recovery

depth, not exceeding 5m	m	50.00
depth, 5-10m	m	60.00
depth, 10-20m	m	70.00
depth, 20-30m	m	80.00
depth, 30-40m	m	90.00

CLASS C: GEOTECHNICAL AND OTHER SPECIALIST PROCESSES

Drilling

An allowance of £3,000 should
be made for the establishment
and removal of plant and equipment
to carry out drilling operations.

£

Drilling for grout holes
through materials other
than rock or artificial
hard materials

vertically downwards			
depth, not exceeding 5m	m	28.00
depth, 5-10m	m	30.00
depth, 10-20m	m	32.00
depth, 20-30m	m	34.00
downwards at an angle of 0-45 degrees to the vertical		
depth, not exceeding 5m	m	28.00
depth, 5-10m	m	30.00
depth, 10-20m	m	32.00
depth, 20-30m	m	34.00
horizontally or downwards at an angle of 0-45 degrees to the horizontal		
depth, not exceeding 5m	m	28.00
depth, 5-10m	m	30.00
depth, 10-20m	m	32.00
depth, 20-30m	m	34.00
upwards at an angle of 0-45 degrees to the horizontal		
depth, not exceeding 5m	m	34.00
depth, 5-10m	m	36.00
depth, 10-20m	m	38.00
depth, 20-30m	m	40.00
upwards at an angle less than 45 degrees to the vertical		
depth, not exceeding 5m	m	34.00
depth, 5-10m	m	36.00
depth, 5-10m	m	38.00
depth, 20-30m	m	40.00

£

Drilling for grout holes
through rock

		£
vertically downwards		
depth, not exceeding 5m	m	26.00
depth, 5-10m	m	28.00
depth, 10-20m	m	30.00
depth, 20-30m	m	32.00
downwards at an angle of		
0-45 degrees to the vertical		
depth, not exceeding 5m	m	26.00
depth, 5-10m	m	28.00
depth, 10-20m	m	30.00
depth, 20-30m	m	32.00
horizontally or downwards		
at an angle of 0-45 degrees to		
the horizontal		
depth, not exceeding 5m	m	26.00
depth, 5-10m	m	28.00
depth, 10-20m	m	30.00
depth, 20-30m	m	32.00
upwards at an angle of 0-45		
degrees to the horizontal		
depth, not exceeding 5m	m	32.00
depth, 5-10m	m	34.00
depth, 10-20m	m	36.00
depth, 20-30m	m	38.00
upwards at an angle less than		
45 degrees to the vertical		
depth, not exceeding 5m	m	32.00
depth, 5-10m	m	34.00
depth, 10-20m	m	36.00
depth, 20-30m	m	38.00

Grout materials and injections   £

An allowance of £3,000 should be made for the
establishment and renewal of plant and equipment
to carry out grouting operations.

Materials

Cement bentonite (2:1)	t	105.00
Cement PFA	t	90.00
Sand	t	18.00
Pea gravel	t	16.00
Bentonite	t	150.00

Injections

Number of holes	nr	80

Diaphragm walls

It is assumed in the following rates that a minimum of 3000 m3
of excavation is required. An allowance of £40,000 should be
made for the establishment and removal of plant and equipment
to carry out the work.

Excavation in material other
than rock or artificial hard
material

maximum depth not exceeding 5m	m3	150.00
maximum depth 5-10m	m3	155.00
maximum depth 10-15m	m3	160.00
maximum depth 15-20m	m3	165.00

		£
Concrete designed mix to BS5328; grade 20, ordinary Portland cement to BS12, 20mm aggregate to BS882, walls 1000mm thick	m3	120.00
High yield bar reinforcement to BS4449		
nominal size 12mm	t	650.00
nominal size 16mm	t	630.00
nominal size 20mm	t	600.00
nominal size 25mm	t	580.00
Concrete guide wall at the side of excavation, 1000mm wide × 500mm deep	m	110.00

Ground anchorages

It is assumed in the following rates that a minimum of 75 anchors are to be installed and an allowance of £4,000 should be made for the establishment and removal of plant and equipment.

Ground anchorages, in material other than rock to a maximum 10m depth; load 50 tonne

temporary	nr	100.00
temporary with single corrosion protection	nr	100.00
temporary with double corrosion protection	nr	100.00
permanent	nr	100.00
permanent with single corrosion protection	nr	100.00
permanent with double corrosion protection	nr	100.00

£

Length of tendon in material other than rock

temporary	m		80.00
temporary with single corrosion protection	m		85.00
temporary with double corrosion protection	m		90.00
permanent	m		95.00
permanent with single corrosion protection	m		100.00
permanent with double corrosion protection	m		105.00

Ground anchorages, in material including
rock to a maximum 10m depth; load 50 tonne

temporary	nr		125.00
temporary with single corrosion protection	nr		125.00
temporary with double corrosion protection	nr		125.00
permanent	nr		125.00
permanent with single corrosion protection	nr		125.00
permanent with double corrosion protection	nr		125.00

Total length of tendon in material including rock

temporary	m		90.00
temporary with single corrosion protection	m		95.00
temporary with double corrosion protection	m		100.00
permanent	m		105.00
permanent with single corrosion protection	m		110.00
permanent with double corrosion protection	m		115.00

Sand, band and wick drains   £

It is assumed in the following rates that a minimum
of 100 vertical drains are to be installed and
an allowance of £6,000 should be made for the
establishment and removal of plant and equipment.

Number of drains	70 nr		
Pre-drilled holes	70 nr		

Drains of maximum depth not
exceeding 10m

cross section		
100-200mm	m	8.00
200-300mm	m	9.00
300-400mm	m	10.00

Drains of maximum depth
10-15m

cross section		
100-200mm	m	9.00
200-300mm	m	10.00
300-400mm	m	11.00

Drains of maximum depth
15-20m

cross section		
100-200mm	m	10.00
200-300mm	m	11.00
300-400mm	m	12.00

Class D: DEMOLITION AND SITE CLEARANCE

General clearance

General site clearance of areas

		£
free from major obstructions	ha	800.00
woods, small trees and shrubs	ha	1500.00

Demolish existing buildings
including digging up
foundations

brick		
small	m3	10.00
medium	m3	8.00
large	m3	4.00
steel framed with cladded walls		
small	m3	4.00
medium	m3	3.00
large	m3	2.00
timber framed with cladded walls		
small	m3	3.00
medium	m3	2.00
large	m3	1.50

Pull down trees (stumps
measured separately) girth

500mm-1m	nr	40.00
1-2m	nr	80.00
2-3m	nr	200.00
3-5m	nr	1250.00

		£

Grub up stumps and backfill
with displaced topsoil, stump
diameter

150-500mm	nr	30.00
500mm-1m	nr	60.00
1-2m	nr	80.00

Dig out and remove drains
depth 1.5m including granular
bed and surround · m · 10.00

Dig up and remove drains depth
2m including concrete bed
and surround · m · 13.00

CLASS E: EARTHWORKS

Dredging

It is difficult to provide accurate cost information on dredging. Vario
methods can be used including cutter suction dredger, barge mounted
excavator or grab hopper. The depth of water, disposal arrangements
and tidal conditions are all key factors affecting costs.

The cost per cubic metre of dredging solid material should
be in the range of £3 to £8 per cubic metre but specialist
advice should be obtained in the early stages of preparing
the budget estimate.

General excavation   £

Excavate to reduce levels

depth not exceeding 0.25m		
topsoil	m3	2.00
normal ground	m3	2.00
stiff clay	m3	3.50
chalk	m3	4.50

		£
depth 0.25 to 0.50m		
normal ground	m3	2.00
stiff clay	m3	3.50
chalk	m3	4.50
rock	m3	12.00
depth 0.50 to 1m		
normal ground	m3	2.00
stiff clay	m3	3.50
chalk	m3	4.50
rock	m3	12.00
depth 1m to 2m		
normal ground	m3	2.50
stiff clay	m3	4.00
chalk	m3	5.50
rock	m3	14.00
depth 2m to 5m		
normal ground	m3	5.00
stiff clay	m3	8.00
chalk	m3	13.00
rock	m3	34.00

Disposal

Excavated material

deposited on site 100m distance	m3	3.00
deposited on site 300m distance	m3	3.50
deposited off site 1km distance including tipping fees	m3	5.00
deposited off site 5km distance including tipping fees	m3	8.00

£

Filling

Filling to make up levels including
levelling and compacting

excavated material	m3	5.00
sand	m3	20.00
hardcore	m3	15.00
DOT Type 1	m3	25.00
DOT Type 2	m3	23.00

CLASS F: IN SITU CONCRETE

Provision of concrete

Standard mix

ST1

ordinary Portland cement	m3	46.00
sulphate resisting cement	m3	50.00

ST2

ordinary Portland cement	m3	48.00
sulphate resisting cement	m3	52.00

ST3

ordinary Portland cement	m3	52.00
sulphate resisting cement	m3	56.00

ST4

ordinary Portland cement	m3	54.00
sulphate resisting cement	m3	58.00

ST5

ordinary Portland cement	m3	58.00
sulphate resisting cement	m3	62.00

Standard mix, ordinary Portland cement £

grade 20

10mm aggregate	m3	64.00
14mm aggregate	m3	62.00
20mm aggregate	m3	60.00
40mm aggregate	m3	58.00

grade 25

10mm aggregate	m3	66.00
14mm aggregate	m3	64.00
20mm aggregate	m3	62.00
40mm aggregate	m3	60.00

grade 30

10mm aggregate	m3	68.00
14mm aggregate	m3	66.00
20mm aggregate	m3	64.00
40mm aggregate	m3	62.00

Designed mix, ordinary Portland cement

grade 20

10mm aggregate	m3	68.00
14mm aggregate	m3	66.00
20mm aggregate	m3	64.00
40mm aggregate	m3	62.00

grade 25

10mm aggregate	m3	70.00
14mm aggregate	m3	68.00
20mm aggregate	m3	66.00
40mm aggregate	m3	64.00

grade 30

10mm aggregate	m3	72.00
14mm aggregate	m3	70.00
20mm aggregate	m3	68.00
40mm aggregate	m3	66.00

Placing of concrete £

Mass concrete in

blinding, thickness		
not exceeding 150mm	m3	13.00
150-300mm	m3	12.00
300-500mm	m3	11.00
exceeding 500mm	m3	10.00
bases, footings, pile caps and ground slabs		
not exceeding 150mm	m3	14.00
150-300mm	m3	13.00
300-500mm	m3	12.00
exceeding 500mm	m3	11.00
walls		
not exceeding 150mm	m3	15.00
150-300mm	m3	14.00
300-500mm	m3	13.00
exceeding 500mm	m3	12.00

Reinforced concrete in

bases, footings, pile caps and ground slabs		
not exceeding 150mm	m3	16.00
150-300mm	m3	15.00
300-500mm	m3	14.00
exceeding 500mm	m3	13.00
walls		
not exceeding 150mm	m3	17.00
150-300mm	m3	16.00
300-500mm	m3	15.00
exceeding 500mm	m3	14.00

		£
suspended slabs		
not exceeding 150mm	m3	20.00
150-300mm	m3	19.00
300-500mm	m3	18.00
exceeding 500mm	m3	17.00
columns and piers		
not exceeding 150mm	m3	38.00
150-300mm	3	36.00
300-500mm	m3	34.00
exceeding 500mm	m3	32.00

CLASS G: CONCRETE ANCILLARIES

Formwork

Rough finish

width 0.2m to 0.4m		
horizontal	m3	32.00
vertical	m3	34.00
curved to radius in one plane	m3	48.00
width exceeding 1.22m		
horizontal	m3	26.00
vertical	m3	28.00
curved to radius in one plane	m3	48.00

Fair finish

width 0.2m to 0.4m		
horizontal	m3	34.00
vertical	m3	36.00
curved to radius in one plane	m3	50.00
width exceeding 1.22m		
horizontal	m3	28.00
vertical	m3	30.00
curved to radius in one plane	m3	44.00

Reinforcement   £

Mild steel bars, diameter

6mm	t	560.00
8mm	t	570.00
10mm	t	580.00
12mm	t	590.00
16mm	t	690.00
20mm	t	760.00
32mm	t	780.00
40mm	t	850.00

High yield steel bars, diameter

6mm	t	860.00
8mm	t	820.00
10mm	t	790.00
12mm	t	750.00
16mm	t	720.00
20mm	t	610.00
25mm	t	590.00
32mm	t	580.00
40mm	t	570.00

Steel fabric, weight

3-4kg/m2	m2	4.50
4-5kg/m2	m2	6.00
5-6kg/m2	m2	8.00
6-7kg/m2	m2	10.00
7-8kg/m2	m2	11.00
8-10kg/m2	m2	12.00

Joints   £

Open surface plain with cork
filler, width 0.5-1m, thickness

10mm	m2	10.00
20mm	m2	15.00
125mm	m2	20.00

Formed surface plain with cork
filler, width 0.5-1m, thickness

10mm	m2	32.00
20mm	m2	34.00
25mm	m2	36.00

CLASS H: PRECAST CONCRETE

Prestressed pre tensioned members

beams		
$100 \times 65 \times 1500$mm	nr	30.00
$200 \times 65 \times 2000$mm	nr	40.00
$250 \times 100 \times 2500$mm	nr	50.00

Copings, sills and weir blocks,
weathered and throated, size

150×75mm	m	15.00
200×75mm	m	20.00
300×75mm	m	25.00

CLASS I: PIPEWORK – PIPES
CLASS J: PIPEWORK – FITTINGS AND VALVES
CLASS K: PIPEWORK – MANHOLES AND PIPEWORK
                           ANCILLARIES
CLASS L: PIPEWORK – SUPPORTS AND PROTECTION

Trench excavation   £

Excavating trenches for pipe
diameter 225mm, backfilling
and removing surplus from site

	trench depth		
	500mm	m	6.00
	1000mm	m	12.00
	1500mm	m	14.00
	2000mm	m	16.00
	2500mm	m	18.00
	3000mm	m	20.00
	3500mm	m	24.00
	4000mm	m	30.00
	4500mm	m	40.00
	5000mm	m	50.00

Excavating trenches for pipe
diameter 300mm, backfilling
and removing surplus from site

	trench depth		
	500mm	m	7.00
	1000mm	m	14.00
	1500mm	m	16.00
	2000mm	m	18.00
	2500mm	m	20.00
	3000mm	m	22.00
	3500mm	m	26.00
	4000mm	m	32.00
	4500mm	m	42.00
	5000mm	m	52.00

£

Excavating trenches for pipe diameter 400mm,
backfilling and removing surplus from site

trench depth		
1500mm	m	30.00
2000mm	m	32.00
2500mm	m	36.00
3000mm	m	40.00
3500mm	m	44.00
4000mm	m	50.00
4500mm	m	60.00
5000mm	m	70.00

Excavating trenches for pipe diameter 525mm,
backfilling and removing surplus from site

trench depth		
1500mm	m	36.00
2000mm	m	40.00
2500mm	m	44.00
3000mm	m	50.00
3500mm	m	60.00
4000mm	m	70.00
4500mm	m	80.00
5000mm	m	90.00

Excavating trenches for pipe diameter 900mm,
backfilling and removing surplus from site

trench depth		
1500mm	m	50.00
2000mm	m	60.00
2500mm	m	74.00
3000mm	m	80.00
3500mm	m	100.00
4000mm	m	120.00
4500mm	m	140.00
5000mm	m	160.00

Beds, haunches and surround   £

Bed of sand 150mm thick to pipe
diameter

225mm	m	4.00
300mm	m	5.00
400mm	m	6.00
525mm	m	7.00
900mm	m	12.00

Bed of granular material
150mm thick to pipe
diameter

225mm	m	5.00
300mm	m	6.00
400mm	m	8.00
525mm	m	10.00
900mm	m	14.00

Bed of concrete 150mm
thick to pipe diameter

225mm	m	16.00
300mm	m	18.00
400mm	m	20.00
525mm	m	22.00
900mm	m	26.00

Bed and haunching of
concrete to pipe diameter

225mm	m	20.00
300mm	m	36.00
400mm	m	44.00
525mm	m	52.00
900mm	m	76.00

£

Bed and surround of sand
to pipe diameter

225mm	m	10.00
300mm	m	14.00
400mm	m	18.00
525mm	m	24.00
900mm	m	36.00

Bed and surround of
granular material
to pipe diameter

225mm	m	12.00
300mm	m	16.00
400mm	m	20.00
525mm	m	28.00
900mm	m	40.00

Bed and surround of concrete
to pipe diameter

225mm	m	30.00
300mm	m	36.00
400mm	m	48.00
525mm	m	60.00
900mm	m	120.00

Pipes laid in prepared trenches

Vitrified clay to BS65, spigot and
socket joints with sealing ring,
nominal bore

225mm	m	46.00
300mm	m	56.00
375mm	m	64.00
450mm	m	130.00

£

Vitrified clay to BS65, plain
ended with sleeve joints, nominal
bore

225mm	m	50.00
300mm	m	68.00

Concrete pipes Class L with
rebated flexible plain ended
joints, nominal bore

225mm	m	40.00
300mm	m	44.00
450mm	m	48.00
525mm	m	54.00
600mm	m	60.00
750mm	m	84.00
900mm	m	120.00
1200mm	m	180.00
1500mm	m	290.00
1800mm	m	360.00

Concrete pipes Class M with
rebated flexible plain ended
joints, nominal bore

225mm	m	40.00
300mm	m	44.00
450mm	m	48.00
525mm	m	54.00
600mm	m	60.00
750mm	m	84.00
900mm	m	120.00
1200mm	m	180.00
1500mm	m	290.00
1800mm	m	360.00

£

**Concrete pipes Class H with
rebated flexible plain ended
joints, nominal bore**

225mm	m	40.00
300mm	m	44.00
450mm	m	48.00
525mm	m	54.00
600mm	m	60.00
750mm	m	84.00
900mm	m	120.00
1200mm	m	180.00
1500mm	m	290.00

**Ductile spun iron pipes
with spigot and socket Tyton
joints, nominal bore**

100mm	m	38.00
150mm	m	54.00
250mm	m	80.00
300mm	m	90.00
400mm	m	110.00
450mm	m	110.00
600mm	m	148.00

**Ductile spun iron pipes
with Stantyte joints,
nominal bore**

800mm	m	200.00
900mm	m	220.00
1000mm	m	280.00
1200mm	m	380.00
1600mm	m	700.00

£

Carbon steel pipes with
welded joints, nominal
bore

100mm	m	40.00
150mm	m	50.00
200mm	m	65.00
250mm	m	90.00
300mm	m	110.00

Unplasticised PVC pipes
with ring seal joints,
nominal bore

50mm	m	18.00
100mm	m	26.00
150mm	m	44.00
200mm	m	50.00
250mm	m	70.00
300mm	m	100.00

Blue MDPE pipes with
butt welded joints, outside
diameter

90mm	m	22.00
125mm	m	28.00
160mm	m	34.00
180mm	m	40.00
250mm	m	50.00
315mm	m	65.00
355mm	m	80.00
400mm	m	90.00

Manholes

£

Precast concrete manholes
complete including excavation
concrete surround, base and
cover slab, channels, step
irons and inspection cover

675mm internal diameter			
	depth, 5m	nr	400.00
	depth, 1m	nr	600.00
	depth, 1.5m	nr	700.00
	depth, 2m	nr	800.00
	depth, 2.5m	nr	900.00
	depth, 3m	nr	1000.00
	depth, 3.5m	nr	1100.00
	depth, 4m	nr	1200.00
900mm internal diameter			
	depth, 1m	nr	800.00
	depth, 1.5m	nr	900.00
	depth, 2m	nr	1000.00
	depth, 2.5m	nr	1100.00
	depth, 3m	nr	1200.00
	depth, 3.5m	nr	1300.00
	depth, 4m	nr	1400.00
	depth, 4.5m	nr	1500.00
	depth, 5m	nr	1700.00
1200mm internal diameter			
	depth, 1.5m	nr	1200.00
	depth, 2m	nr	1350.00
	depth, 2.5m	nr	1500.00
	depth, 3m	nr	1650.00
	depth, 3.5m	nr	1800.00
	depth, 4m	nr	1950.00
	depth, 4.5m	nr	2100.00
	depth, 5m	nr	2250.00

£

1500mm internal diameter

depth, 2m	nr	1650.00
depth, 2.5m	nr	1800.00
depth, 3m	nr	1950.00
depth, 3.5m	nr	2100.00
depth, 4m	nr	2250.00
depth, 4.5m	nr	2400.00
depth, 5m	nr	2550.00

1800mm internal diameter

depth, 2m	nr	1900.00
depth, 2.5m	nr	2200.00
depth, 3m	nr	2500.00
depth, 3.5m	nr	2800.00
depth, 4m	nr	3100.00
depth, 4.5m	nr	3400.00
depth, 5m	nr	3700.00

Gullies

Vitrified clay road gully,
480mm diameter × 900 mm
deep, including excavation,
concrete surround, engineering
brick seating and cast iron
grating   nr   540.00

Precast concrete road gully,
375mm diameter × 900mm deep,
including excavation, concrete
surround, Class B engineering
brick seating and cast iron grating   nr   200.00

CLASS M: STRUCTURAL METALWORK   £

Fabrication of members for frames

columns	t	800.00
beams	t	850.00
portal frames	t	900.00
trusses and built up girders	t	1250.00
bracings, purlins and cladding rails	t	1300.00
anchorages and holding down assemblies	t	1450.00

Permanent erection of members for frames	t	200.00

Site bolts

black	t	2400.00
HSFG general grade	t	2400.00
HSFG higher grade	t	2800.00
HSFG load indicating or load limit	t	3400.00

Offsite surface treatment

blast cleaning	m2	3.00
wire brushing	m2	2.00
galvanizing	m2	12.00
one coat chromate primer	m2	3.00

CLASS N: MISCELLANEOUS METALWORK

Stairways and landings	t	2750.00
Walkways and platforms	t	2500.00

		£
Cat ladder in galvanized steel rungs at 300mm centres, strings extended to form handrail, 450mm wide, length		
3m	nr	750.00
4m	nr	1000.00
5m	nr	1250.00
6m	nr	1500.00
7m	nr	1750.00
8m	nr	2000.00
9m	nr	2250.00
10m	nr	2500.00
Guard cage to cat ladder	m	50.00
Galvanized steel staircase 900mm wide with chequer plate treads balustrade one side, supported on universal columns		
5500mm going, 3000mm rise, 16 treads and one landing	nr	2650.00
10000mm going, 5000mm rise, 16 treads and two landings	nr	3750.00
Galvanized tubular handrail, 1050mm high, standards at 2000mm centres, with middle rail	m	100.00
Galvanized flat section handrail and members, standards at 1000mm centres, infilled with square vertical bars at 100mm centres	m	125.00

Safety fencing   £

Tensioned corrugated to DTp
Clause 409 with 'Z' section steel
posts set in concrete

single side	m	55.00
double sided	m	65.00

Untensioned corrugated to
DTp Clause 412, single sided
with timber posts · m · 25.00

Miscellaneous framing

Angle section

150 × 75 × 10mm	m	20.00
100 × 100 × 10mm	m	20.00
150 × 150 × 12mm	m	25.00

Channel section

150 × 75 × 10mm	m	20.00
250 × 75 × 16mm	m	30.00

Flooring

Galvanized mild steel 'Durbar'
pattern plate 8mm thick · m2 · 100.00

Galvanized open grid flooring
50mm thick · m2 · 60.00

CLASS O: TIMBER   £

Greenheart timber in
marine works

100 × 75mm	m	18.00
150 × 75mm	m	24.00
200 × 200mm	m	50.00
200 × 300mm	m	70.00
300 × 300mm	m	90.00
600 × 600mm	m	250.00

Wrought softwood in
marine work

100 × 75mm	m	15.00
150 × 75mm	m	18.00
200 × 200mm	m	40.00
200 × 300mm	m	65.00
300 × 300mm	m	75.00
600 × 600mm	m	200.00

Hardwood decking,
thickness

50mm	m2	70.00
75mm	m2	90.00
100mm	m2	120.00

Softwood decking,
thickness

50mm	m2	50.00
75mm	m2	70.00
100mm	m2	100.00

Metalwork   £

Coach screws, 10mm diameter, length

75mm		2.00
100mm	nr	3.00
150mm	nr	4.00

Blackbolts, nuts and washers, M12, length

100mm	nr	3.00
140mm	nr	4.00
200mm	nr	5.00

CLASS P: PILING
CLASS Q: PILING ANCILLARIES

Bored cast-in-place concrete piles

Allow £6,000 and £10,000 for the cost of setting up
and removing from site for 50 piles and 100 piles respectively.

Reinforced in situ concrete
piles 300mm diameter

concrete	m	25.00
depth bored, 10m	m	22.00
depth bored, 15m	m	24.00
depth bored, 20m	m	26.00
depth bored, 25m	m	28.00

Reinforced in situ concrete
piles 600mm diameter

concrete	m	60.00
depth bored, 10m	m	64.00
depth bored, 15m	m	66.00
depth bored, 20m	m	68.00
depth bored, 25m	m	70.00

£

Reinforced in situ concrete
piles 900mm diameter

concrete	m	80.00
depth bored, 10m	m	90.00
depth bored, 15m	m	95.00
depth bored, 20m	m	100.00
depth bored, 25m	m	105.00

Reinforced in situ concrete
piles 1200mm diameter

concrete	m	120.00
depth bored, 10m	m	160.00
depth bored, 15m	m	170.00
depth bored, 20m	m	180.00
depth bored, 25m	m	190.00

Driven cast-in-place concrete piles

Allow £3,500 and £7,000 for the cost of setting up
and removing from site for 50 piles and 100 piles respectively.

Driven concrete piles
300mm diameter

depth driven, 10m	m	36.00
depth driven, 15m	m	38.00
depth driven, 20m	m	40.00
depth driven, 25m	m	42.00

Driven concrete piles
600mm diameter

depth driven, 10m	m	42.00
depth driven, 15m	m	44.00
depth driven, 20m	m	46.00
depth driven, 25m	m	48.00

£

Driven concrete piles
900mm diameter

depth driven, 10m	m	54.00
depth driven, 15m	m	56.00
depth driven, 20m	m	58.00
depth driven, 25m	m	60.00

Steel sheet piling

Frodingham steel piles,
ref.1N, driven into ground   m2   80.00

extra for corner sections	m	70.00
extra for junctions	m	90.00

Frodingham steel piles,
ref.2N, driven into ground   m2   90.00

extra for corner sections	m	70.00
e×xra for junctions	m	90.00

Frodingham steel piles,
ref.3N, driven into ground   m2   100.00

extra for corner sections	m	70.00
extra for junctions	m	90.00

Frodingham steel piles,
ref.4N, driven into ground   m2   110.00

extra for corner sections	m	70.00
extra for junctions	m	90.00

CLASS R: ROADS AND PAVINGS   £

Sub-bases flexible road bases and surfacing

Hardcore road base	m3	18.00
depth, 100mm	m2	2.00
depth, 150mm	m2	3.00
depth, 200mm	m2	4.00
depth, 250mm	m2	5.00
Granular material, DTp type 1	m3	20.00
depth, 100mm	m2	3.00
depth, 150mm	m2	4.00
depth, 200mm	m2	5.00
depth, 250mm	m2	6.00
Granular material, DTp type 2	m3	20.00
depth, 100mm	m2	3.00
depth, 150mm	m2	4.00
depth, 200mm	m2	5.00
depth, 250mm	m2	6.00
Dense bitumen macadam, DTp clause 908, 14mm aggregate, wearing course, depth		
30mm	m2	3.00
40mm	m2	4.00
Dense bitumen macadam, DTp clause 908, 40mm aggregate, wearing course, depth		
60mm	m2	6.00
100mm	m2	8.00

Concrete pavements   £

Carriageway slabs, concrete
grade C20, depth

150mm	m2	15.00
200mm	m2	18.00
350mm	m2	21.00
200mm	m2	24.00

Steel fabric reinforcement
to BS4483

reference A142 2.22 kgs/m2	m2	4.00
reference B503 5.93 kgs/m2	m2	7.00

Waterproof membranes
below concrete pavements,
plastic sheeting, 1200 gauge   m2   2.00

Joints in concrete pavements

Longitudinal joints, 10mm
diameter × 750mm long mild steel
dowels at 750mm centres, sealed
with polysulphide, depth

150mm	m	25.00
220mm	m	30.00
250mm	m	35.00

Kerbs, channels and edgings

Precast concrete kerbs to BS340
straight or curved to radius
e×ceeding 12m

fig. 6 150 × 305mm	m	16.00
fig. 7 125 × 255mm	m	12.00

CLASS S: RAIL TRACK   £

Track foundations

Bottom ballast granite, crushed graded 50-25mm	m3	34.00
Top ballast granite, crushed, graded 50-25mm	m3	36.00

Taking up track and turnouts, dismantle and stack

Bullhead or flat bottom rail

plain track, timber sleepers, fishplate joints	m	10.00
turnouts, timber sleepers, fishplate joints	nr	500.00
plain track, concrete sleepers, fishplate joints	m	12.00
turnouts, concrete sleepers, fishplate joints	nr	550.00

Lifting packing and slewing

Bullhead rail track on timber sleepers track length 10m, maximum slew 300mm, maximum lift 100mm	nr	400.00

Supplying only plain line material

Bullhead rails; for joints or welded track

mass 40-50 kg/m, section reference 95R	t	650.00
Sleepers, softwood timber, 250 × 125 × 2600mm long	nr	34.00

		£
Sleepers, hardwood timber, 250 × 125 × 2600mm long	nr	40.00
Sleepers, concrete type 'F27' with 2 nr cast iron 'Pandrol' fittings cast in	nr	55.00

Fittings

Chairs, cast iron 'CC' pattern	nr	32.00
Fish plates, standard set	nr	20.00
Fish plates, insulated set	nr	105.00
Switches and crossings		
turnouts	nr	15,000.00
diamond crossings	nr	30,000.00
Buffer stops, 2-2.5 tonnes	nr	2,000.00

Laying only plain line material

Bullhead rails

plain track, mass 40-50kg/m fish plate joints, timber sleepers	m	40.00
welded joints, concrete sleepers	m	45.00
turnout, standard type on timber sleepers	nr	1750.00
buffer stop, single rake	nr	200.00

CLASS T: TUNNELS   £

There are many different methods of boring and
constructing tunnels so the following rates should
be treated with caution. For costs on specific jobs the
advice of a tunnelling contractor should be sought.

Excavation

Tunnels in rock, straight

diameter 1.8m	m3	350.00
diameter 3.0m	m3	240.00

Tunnels in clay, straight

diameter 1.8m	m3	180.00
diameter 3.0m	m3	110.00

Shafts in rock, vertical

diameter 3.0m	m3	140.00
diameter 4.5m	m3	100.00

Shafts in clay, vertical

diameter 3.0m	m3	90.00
diameter 4.5m	m3	70.00

In situ lining to tunnels, vertical

Cast concrete grade C20, diameter 2m	m3	170.00

In situ lining to shafts, vertical

Cast concrete grade C20, diameter 2m	m3	140.00

		£
Formwork, rough finish		
diameter 2m	m2	30.00
Formwork, smooth finish		
diameter 2m	m2	34.00

Pre-formed segmental linings to tunnels

Precast concrete bolted
flanged rings, depth 450mm

diameter 2.5m	nr	300.00
diameter 3.m	nr	650.00

Pre-formed segmental linings to shafts

Precast concrete bolted flanged
rings, depth 450mm

diameter 3m	nr	500.00
diameter 4.5m	nr	1000.00

CLASS U: BRICKWORK, BLOCKWORK AND MASONRY

Common brickwork

Brickwork (£120 per 1000) in
cement mortar (1:3)

vertical walls		
102.5mm thick	m2	30.00
215mm thick	m2	50.00
327mm thick	m2	75.00
440mm thick	m2	90.00

		£

In columns and piers

215m × 215mm	m	25.00
327 × 327mm	m	30.00
440 × 440m	m	35.00

Facing brickwork

Brickwork (£300 per 1000) in
gauged mortar (1:1:6)

vertical walls		
102.5mm thick	m2	40.00
215mm thick	m2	70.00

Engineering brickwork (£200 per
1000) in cement mortar (1:3)

vertical walls		
215mm thick	m2	35.00
327mm thick	m2	60.00
440mm thick	m2	85.00

Engineering brickwork, (£250 per
1000) in cement mortar

facing to vertical walls		
102.5mm thick	m2	38.00
215mm thick	m2	65.00

Lightweight blockwork

Blockwork in cement
mortar (1:3) in vertical
straight walls

100mm thick	m2	18.00
140mm thick	m2	22.00
190mm thick	m2	30.00

Dense concrete blockwork   **£**

Blockwork in cement
mortar (1:3) in vertical
straight walls

100mm thick	m2	22.00
140mm thick	m2	28.00
190mm thick	m2	36.00

Ashlar masonry

Portland Whitbed with one
exposed face in gauged mortar
(1:1:6)

facing to vertical walls

50mm thick	m2	200.00
75mm thick	m2	250.00
100mm thick	m2	300.00
150mm thick	m2	400.00
200mm thick	m2	600.00

CLASS V: PAINTING
(Rates inclusive of all inclinations)

One coat primer on general
surfaces exceeding 300mm

metal	m2	2.00
timber	m2	2.00

Two coats emulsion paint
on general surfaces
exceeding 300mm

smooth concrete	m2	3.00
blockwork and brickwork	m2	4.00

		£
Two coats of cement paint		
smooth concrete	m2	3.00
blockwork and brickwork	m2	4.00
rough cast surfaces	m2	5.00
Three coats of oil paint on		
primed steel sections	m2	7.00
primed pipework	m2	8.00
planed timber	m2	7.00

CLASS W: WATERPROOFING

Damp proofing

One layer 1000 gauge 'Bituthene' sheet, fixed with adhesive	m2	8.00
Asphalt to BS1097 two coat work, 20mm thick on concrete surfaces	m2	30.00

Roofing

Asphalt to BS908, two coat work, 20mm thick on concrete surfaces	m2	30.00
Built up felt roofing to BS747, three layer coverings	m2	25.00

Waterproofing

Proprietary roof decking with roof felt finish, insulated	m2	40.00

£

Protective layers, one layer 1000 gauge polythene sheet, fixed with adhesive	m2	2.00
Cement and sand (1:3) screed with waterproof additive	m2	10.00
Sprayed or brushed waterproofing two coats of 'Synthaprufe' to concrete surfaces	m2	5.00

CLASS X: MISCELLANEOUS WORK

See Landscaping section

CLASS Y: SEWER RENOVATION AND ANCILLARY WORKS

Preparation of existing sewers

Cleaning egg shaped sewer 1050mm high	m	15.00

Removing intrusions

brickwork	m3	75.00
concrete	m3	100.00

Plugging laterals with concrete

bore not exceeding 300mm	nr	75.00
bore exceeding 300mm	nr	125.00

Local internal repairs

areas 0.1-0.25m2	nr	50.00
area 5m2	nr	150.00

Stabilisation of existing sewers		£
Pointing with cement mortar (1:3)	m2	30.00

Renovation of existing sewers

Segmental lining in GRP

egg shaped 1050mm high	m	300.00
annulus grouting in Pozament	m3	225.00

Laterals to renovated sewers

Jointing

bore not e×ceeding 150mm	nr	60.00
bore 150-300m	nr	110.00
bore 450mm	nr	170.00

Interruptions

Preparation of existing sewers

cleaning	hour	300.00

Stabilisation

pointing	hour	75.00

Renovation

linings	hour	80.00

CLASS Z: SIMPLE BUILDING WORKS INCIDENTAL TO CIVIL ENGINEERING WORKS

See Building work

MECHANICAL WORK

Pipework   £

Mild steel pipes, heavy, black,
malleable iron, fixing with
standard supports, including
fittings

screwed fittings, pipe size		
20mm	m	25.00
25mm	m	28.00
32mm	m	31.00
40mm	m	34.00
50mm	m	56.00
65mm	m	72.00
80mm	m	80.00
100mm	m	90.00
125mm	m	100.00
150mm	m	140.00
welded fittings, pipe size		
20mm	m	30.00
25mm	m	34.00
32mm	m	38.00
40mm	m	44.00
50mm	m	60.00
65mm	m	80.00
80mm	m	90.00
100mm	m	100.00
125mm	m	110.00
150mm	m	150.00

£

Mild steel pipes, medium, black, malleable iron,
fixing with standard supports, including fittings

    screwed fittings, pipe size

20mm	m	22.00
25mm	m	26.00
32mm	m	28.00
40mm	m	32.00
50mm	m	50.00
65mm	m	66.00
80mm	m	74.00
100mm	m	84.00
125mm	m	94.00
150mm	m	130.00

    welded fittings, pipe size

20mm	m	24.00
25mm	m	32.00
32mm	m	36.00
40mm	m	42.00
50mm	m	56.00
65mm	m	74.00
80mm	m	84.00
100mm	m	94.00
125mm	m	104.00
150mm	m	140.00

Copper pipes; BS2871 part 1 table X, fixing
with standard supports, including fittings

    capillary fittings, pipe size

15mm	m	24.00
22mm	m	26.00
28mm	m	28.00
35mm	m	32.00
42mm	m	44.00
54mm	m	50.00

		£
compression fittings, pipe size		
15mm	m	28.00
22mm	m	30.00
28mm	m	32.00
35mm	m	36.00
42mm	m	48.00
54mm	m	54.00

Carbon steel pipes to BS3601, fixing with
standard supports, including fittings

welded fittings, pipe size		
15mm	m	28.00
20mm	m	32.00
25mm	m	38.00
32mm	m	44.00
40mm	m	48.00
50mm	m	58.00
65mm	m	72.00
80mm	m	88.00
100mm	m	116.00
125mm	m	160.00
150mm	m	190.00

Ductwork

Rectangular galvanised sheet
steel ductwork including
fittings, supports and doors

girth of 2 sides		
250mm	m	40.00
500mm	m	46.00
750mm	m	56.00
1000mm	m	70.00
1250mm	m	80.00
1500mm	m	86.00
1750mm	m	96.00
2000mm	m	110.00

£

2250mm	m	136.00
2500mm	m	150.00
2750mm	m	180.00
3000mm	m	200.00

Circular galvanised steel
spirally wound ductwork
including fittings, supports
and doors, diameter

100mm	m	24.00
150mm	m	28.00
200mm	m	34.00
250mm	m	46.00
300mm	m	52.00
350mm	m	58.00
400mm	m	66.00
450mm	m	74.00
500mm	m	80.00
550mm	m	86.00
600mm	m	92.00
650mm	m	98.00
700mm	m	104.00
750mm	m	112.00
800mm	m	120.00
850mm	m	130.00
900mm	m	134.00
950mm	m	142.00
1000mm	m	154.00

Thermal insulation

Rigid mineral wool sections,
bright class 'O' foil covered,
secured with aluminium bands at
300mm centres, including for
fixing around joints, flanges,
valves and the like

£

insulation thickness 25mm,
pipe size

15mm	m	5.00
20mm	m	7.00
25mm	m	8.00
32mm	m	10.00
40mm	m	12.00
50mm	m	18.00
65mm	m	24.00
80mm	m	36.00
100mm	m	42.00

insulation thickness 40mm,
pipe size

15mm	m	6.00
20mm	m	8.00
25mm	m	10.00
32mm	m	12.00
40mm	m	14.00
50mm	m	20.00
65mm	m	28.00
80mm	m	40.00
100mm	m	44.00

insulation thickness 50mm,
pipe size

15mm	m	12.00
20mm	m	14.00
25mm	m	18.00
32mm	m	22.00
40mm	m	24.00
50mm	m	28.00
65mm	m	30.00
80mm	m	40.00
100mm	m	50.00

£

Boilers

Domestic gas fired central heating boilers,
floor mounted, balanced flue

30,000 Btu	nr	800.00
40,000 Btu	nr	850.00
50,000 Btu	nr	930.00
60,000 Btu	nr	1000.00
70,000 Btu	nr	1100.00
80,000 Btu	nr	1200.00
90,000 Btu	nr	1350.00
100,000 Btu	nr	1400.00

Domestic gas fired central heating boilers,
wall mounted, balanced flue

30,000 Btu	nr	850.00
40,000 Btu	nr	900.00
50,000 Btu	nr	980.00
60,000 Btu	nr	1050.00
70,000 Btu	nr	1150.00
80,000 Btu	nr	1250.00
90,000 Btu	nr	1350.00
100,000 Btu	nr	1450.00

Commercial packaged low pressure hot
water gas fired boiler complete

50kw	nr	2000.00
100kw	nr	3000.00
150kw	nr	5500.00
200kw	nr	6500.00
250kw	nr	7000.00
300kw	nr	8000.00
350kw	nr	8500.00
450kw	nr	1150.00
600kw	nr	1400.00

£

1500kw	nr	2000.00
2000kw	nr	3000.00
3000kw	nr	4000.00

Commercial packaged
low pressure hot water
oil fired boiler complete

50kw	nr	1500.00
100kw	nr	2800.00
150kw	nr	5000.00
200kw	nr	6000.00
250kw	nr	6500.00
300kw	nr	7000.00
350kw	nr	8000.00
450kw	nr	1100.00
600kw	nr	1300.00
1500kw	nr	1800.00
2000kw	nr	2800.00
3000kw	nr	3700.00

Radiators

Pressed steel single panel
radiator 450mm high complete
with valves, length

400mm	nr	80.00
600mm	nr	90.00
800mm	nr	100.00
1000mm	nr	11500
1200mm	nr	12500
1400mm	nr	135.00
1600mm	nr	150.00
1800mm	nr	160.00
2000mm	nr	170.00

£

Pressed steel single panel
radiator 700mm high complete
with valves, length

400mm	nr	90.00
600mm	nr	100.00
800mm	nr	110.00
1000mm	nr	125.00
1200mm	nr	135.00
1600mm	nr	160.00
1800mm	nr	170.00
2000mm	nr	180.00

Pressed steel double panel
radiator 450mm high complete
with valves, length

400mm	nr	100.00
600mm	nr	110.00
800mm	nr	120.00
1000mm	nr	135.00
1200mm	nr	145.00
1600mm	nr	180.00
1800mm	nr	190.00
2000mm	nr	2000.00

Pressed steel double panel
radiator 700mm high complete
with valves, length

400mm	nr	110.00
600mm	nr	120.00
800mm	nr	130.00
1000mm	nr	145 00
1200mm	nr	155.00
1400mm	nr	165.00
1600mm	nr	190.00
1800mm	nr	200.00
2000mm	nr	210.00

ELECTRICAL WORK

Transformers   £

Transformer, 11kV/415 volt,
50 Hz three phase, air cooled,
oil filled, skid mounted, cable
boxes, fixing to backgrounds

500 kVA	nr	6000.00
800 kVA	nr	7000.00
1000 kVA	nr	8000.00
1250 kVA	nr	9500.00
1500 kVA	nr	11000.00
2000 kVA	nr	15000.00

Transformer, 11kV/415 volt,
50 Hz three phase, hermetically
sealed, silicone impregnated, skid
mounted, cable boxes, fixing to backgrounds

500 kVA	nr	7500.00
800 kVA	nr	9500.00
1000 kVA	nr	11500.00
1250 kVA	nr	12500.00
1500 kVA	nr	14000.00
2000 kVA	nr	1,000.00

Sundries

Extra for

conservator	nr	550.00
dial thermometer	nr	400.00
Bucholz relay	nr	380.00

Distribution boards   £

Low voltage distribution boards
steel casing, 500V, 2-20Amp
fuses including fuse links

SP&N

4 way	nr	140.00
6 way	nr	170.00
8 way	nr	200.00
12 way	nr	230.00

TP&N

4 way	nr	220.00
6 way	nr	280.00
8 way	nr	350.00
12 way	nr	430.00

Low voltage switchboards in
steel casing, 500V, 2-30Amp
fuses including fuse links

SP & N

4 way	nr	180.00
6 way	nr	210.00
8 way	nr	240.00
12 way	nr	280.00

TP & N

4 way	nr	500.00
6 way	nr	650.00
8 way	nr	750.00
12 way	nr	950.00

Circuit breakers   £

Non-automatic circuit breakers,
fixed manual

    triple pole

800 Amp	nr	1800.00
1000 Amp	nr	1850.00
1250 Amp	nr	1900.00
1600 Amp	nr	2100.00
2000 Amp	nr	2800.00
2500 Amp	nr	3400.00

    four pole

800 Amp	nr	2200.00
1000 Amp	nr	2250.00
1250 Amp	nr	2500.00
1600Amp	nr	2700.00
2000 Amp	nr	3300.00
2500 Amp	nr	4500.00

Non-automatic circuit breakers,
drawout manual

    triple pole

800 Amp	nr	2500.00
1000 Amp	nr	2600.00
1250 Amp	nr	2900.00
1600 Amp	nr	3000.00
2000 Amp	nr	3750.00
2500 Amp	nr	4800.00

    four pole

800 Amp	nr	2950.00
1000 Amp	nr	3100.00
1250 Amp	nr	3700.00
1600 Amp	nr	3800.00
2000 Amp	nr	4750.00
2500 Amp	nr	6000.00

Busbar trunking   £

Busbar trunking; 4nr copper
conductor bars, extruded
aluminium housing, earth
continuity, fixing to supports

380 amp	m	200.00
end cap	nr	40.00
end feed unit	nr	300.00
centre feed unit	nr	290.00
fire barrier	nr	75.00
440 amp	m	215.00
end cap	nr	45.00
end feed unit	nr	240.00
centre feed unit	nr	230.00
fire barrier	nr	85.00
750 amp	m	260.00
end cap	nr	50.00
tee flat edge	nr	430.00
end feed unit	nr	410.00
fire barrier	nr	95.00

HV/LV cables and wiring

PVC sheathed cable, 6350/11000 volt
grade, lead-covered, paper-insulated single
wire armoured, with copper conductors

single core cable,clipped to backgrounds		
$120mm^2$	m	28.00
$150mm^2$	m	30.00
$185mm^2$	m	34.00
$240mm^2$	m	38.00
$300mm^2$	m	42.00

		£
triple core cable,clipped to backgrounds		
120mm^2	m	42.00
150mm^2	m	46.00
185mm^2	m	50.00
240mm^2	m	60.00
300mm^2	m	70.00

Copper sheathed light duty
mineral insulated cable
500 Volt grade with copper
conductors

sheathed		
2L 1.0	m	5.00
2L 4.0	m	8.00
3L 1.5	m	6.00
3L 2.5	m	7.00
4L 1.5	m	7.00
4L 2.5	m	9.00
7L 1.5	m	10.00
7L 2.5	m	11.00

bare		
2L 1.0	m	4.00
2L 4.0	m	7.00
3L 1.5	m	5.00
3L 2.5	m	6.00
4L 1.5	m	6.00
4L 2.5	m	8.00
7L 1.5	m	9.00
7L 2.5	m	10.00

Cable trays   £

Galvanised steel, heavy duty perforated
cable trays including supports, fixings,
bends, tees and reducers

50mm wide	m	12.00
100mm wide	m	15.00
150mm wide	m	18.00
300mm wide	m	30.00
600mm wide	m	60.00

Conduits

Heavy duty welded steel conduit including
all fittings and supports

galvanised

20mm	m	13.00
25mm	m	14.00
32mm	m	16.00
38mm	m	18.00
50mm	m	28.00

black enamelled

20mm	m	12.00
25mm	m	13.00
32mm	m	15.00
38mm	m	17.00
50mm	m	26.00

Heavy duty galvanised steel
PVC covered conduit
including all fittings and supports

20mm	m	16.00
25mm	m	18.00
32mm	m	20.00
38mm	m	22.00
50mm	m	28.00

Trunkings

£

Galvanised steel trunking,
complete with fittings, fixing
to backgrounds

single compartment			
50 × 50mm	m	18.00
75 × 50mm	m	22.00
75 × 75mm	m	24.00
100 × 50mm	m	25.00
100 × 75mm	m	28.00
100 × 100mm	m	30.00
150 × 50mm	m	34.00
150 × 75mm	m	36.00
150 × 100mm	m	38.00
150 × 150mm	m	44.00
225 × 150mm	m	54.00
225 × 225mm	m	60.00
300 × 225mm	m	70.00
300 × 300mm	m	84.00
twin compartment			
50 × 50mm	m	22.00
75 × 50mm	m	24.00
75 × 75mm	m	26.00
100 × 50mm	m	28.00
100 × 75mm	m	32.00
100 × 100mm	m	34.00
150 × 50mm	m	36.00
150 × 75mm	m	38.00
150 × 100mm	m	40.00
150 × 150mm	m	50.00
225 × 150mm	m	60.00
225 × 225mm	m	70.00
300 × 225mm	m	80.00
300 × 300mm	m	90.00

£

Luminaires

Fluorescent batten type luminaire, single
tube, switch start, fixing to backgrounds

1200mm, 36 watt	nr	56.00
1500mm, 58 watt	nr	60.00
1800mm, 70 watt	nr	66.00
2400mm, 100 watt	nr	78.00

Fluorescent batten type luminaire,
twin tube, switch start,
fixing to backgrounds

1200mm, 36 watt	nr	76.00
1500mm, 58 watt	nr	80.00
1800mm, 70 watt	nr	86.00
2400mm, 100 watt	nr	100.00

Fluorescent batten type luminaire,
twin tube, HF electronic ballast start,
fixing to backgrounds

1200mm, 2 × 36 watt	nr	78.00
1500mm, 2 × 58 watt	nr	85.00
1800mm, 2 × 70 watt	nr	95.00

Road lighting

High pressure sodium lantern,
aluminium canopy, support,
clear bowl, lighting column,
aluminium outreach bracket,
fixing to base

250 watt 10 metre high	nr	1200.00
400 watt 10 metre high	nr	1300.00
250 watt 12 metre high	nr	1500.00
400 watt 12 metre high	nr	1600.00

Access control

		£
Card access door system complete with control processor, monitor, electro magnetic locks and card reader	nr	3500.00
Door hold system complete with power supply unit and electro magnetic locks	nr	500.00

Security detection and alarm

Door and window monitoring complete with control panel/mimic, alarm, transmitter units, reed switches and cable	nr	120.00

Escalators

30 degrees incline, 3-4m rise, glass balustrades and 600 mm steps	nr	64000.00
Extra for 5-6m rise	nr	7000.00
Extra for stainless steel balustrades and decking	nr	8000.00
E×tra for under handrail lighting	nr	4000.00
30 degrees incline, 4-5m rise, glass balustrades and 800 mm steps	nr	70000.00
Extra for 5-6m rise	nr	8000.00
Extra for stainless steel balustrades and decking	nr	9000.00
Extra for under handrail lighting	nr	5000 00

ALTERATIONS AND REPAIRS

Taking down   £

Take down walls and remove
rubble from site

e×ternal walls			
	half brick wall	m2	12.00
	one brick wall	m2	20.00
	one and a half brick wall	m2	26.00
	two brick wall	m2	32.00
	blockwork 140mm wall	m2	17.00
	blockwork 190mm wall	m2	20.00
	blockwork 215mm wall	m2	22.00
internal walls			
	half brick wall	m2	13.00
	one brick wall	m2	22.00
	blockwork 90mm wall	m2	10.00
	blockwork 100mm wall	m2	12.00
	blockwork 115mm wall	m2	13.00
	blockwork 125mm wall	m2	15.00
	blockwork 140mm wall	m2	18.00

Take down and remove chimney
stack to below roof slope, size

750 × 750 × 1200mm above roof	nr	300.00
1030 × 1030 × 1000mm above roof	nr	450.00

Take down chimney breast from
roof to ground level (2 storey)   nr   1600.00

Alterations   £

Form openings in walls

openings not exceeding 0.5m2		
half brick wall	m2	40.00
one brick wall	m2	60.00
brick/block cavity wall	m2	60.00
100mm block wall	m2	30.00
215mm block wall	m2	38.00
openings 0.5-1m2		
half brick wall	m2	30.00
one brick wall	m2	50.00
brick/block cavity wall	m2	50.00
100mm block wall	m2	25.00
215mm block wall	m2	30.00

Cutting opening in internal wall
900 × 2100mm suitable for new single
door including squaring up jambs and
head, inserting lintel over, making
good to surrounding plasterwork

half brick wall	nr	350.00
one brick wall	nr	500.00
one and a half brick wall	nr	620.00
100mm block wall	nr	300.00
215mm block wall	nr	380.00

Cutting opening in external wall 1200 ×
1200mm for new window including
squaring up jambs, head and cill,
inserting lintel over, making good
surrounding plasterwork

one brick wall	nr	450.00
brick/block cavity wall	nr	450.00

£

Filling in openings in common
brickwork in gauged mortar,
including plastering one side

one brick wall	nr	80.00
brick/block cavity wall	nr	80.00

Filling in openings in blockwork
in gauged mortar, including
plastering one side

100mm thick blockwork wall	m2	60.00
215mm thick blockwork wall	m2	75.00

Fill openings in concrete floor with
concrete

100mm thick	m2	12.00
150mm thick	m2	16.00
200mm thick	m2	20.00
250mm thick	m2	24.00

Fill openings in suspended concrete
floor with concrete including formwork
to sofffit

100mm thick	m2	46.00
150mm thick	m2	50.00
200mm thick	m2	54.00
250mm thick	m2	58.00

Cut out defective facing bricks and renew

£350 per thousand

single brick	nr	10.00
areas less than 1m2	m2	60.00
areas over 1m2	m2	50.00

		£
£500 per thousand		
single brick	nr	10.00
areas less than 1m2	m2	70.00
areas over 1m2	m2	60.00

Cut out raking crack in brickwork,
stitch in new facing brickwork
(£350 per thousand) pointing
one side to match existing

half brick thick wall	m	30.00
one brick thick wall	m	50.00

Renew concrete lintel 1500mm long
including one course of brickwork, inserting
new concrete lintel and making good   nr   120.00

Rake out joints in brickwork and
repoint in gauged mortar   m2   20.00

Cut out decayed structural softwood
timber member and renew

floors		
50 × 150mm	m	10.00
50 × 175mm	m	12.00
50 × 200mm	m	14.00

floors		
50 × 150mm	m	12.00
50 × 175mm	m	14.00
50 × 200mm	m	16.00

Remove damaged tile/slate and
replace to match existing

Welsh slate	nr	14.00
concrete interlocking tile	nr	9.00
clay plain tile	nr	8.00

		£
Remove damaged tiles/slates and replace to match existing		
Welsh slates	m2	80.00
concrete interlocking tile	m2	36.00
clay plain tiles	m2	34.00
Cut out defective bituminous felt roofing and renew		
areas less than 1m2	m2	40.00
areas over 1m2	m2	30.00
Take up existing flooring and renew		
chipboard flooring	m2	16.00
tongues and grooved softwood	m2	38.00
Take up softwood skirting and renew	m	7.00
Ease adjust and re-hang single door	nr	10.00
Ease opening casement or sash window	nr	8.00
Cut out decayed transome of sash window and piece in new section	nr	20.00
Repairing sash window, easing, re-cording, replacing weights beads and ironmongery	nr	50.00
Take out timber staircase (single storey height) and fill in opening in timber floor	nr	450.00
Renew tread or riser to timber staircase	nr	28.00
Take up defective stair nosing and replace with aluminium nosing	nr	22.00

		£
Clean out rainwater gutters	m	2.00
Replace galvanised metal support bracket to gutter	nr	5.00
Take down damaged section of Rainwater gutter and replace with new half round PVC-U gutter	m	18.00
Hack off addled plaster and renew		
areas less than 1m2	m2	22.00
areas over 1m2	m2	20.00
Cut out crack to plastered wall and make good	m	5.00
Cut out crack to plastered ceiling and make good	m	5.00
Repair defective floor screed		
areas less than 1m2	m2	14.00
areas over 1m2	m2	12.00
Prepare and refix loose ceramic wall tiles	m2	18.00
Strip off wall coverings, rub down plastered walls and prepare for new finish	m2	2.00

		£
Clean off existing Artex from ceiling and prepare to receive new finish	m2	5.00
Clean down painted timber frames and prepare to receive new finish	m2	1.00
Burn off paint from timber frames, rub down and prepare for new finish	m2	4.00

5

Indices and regional variations

INDICES

Indices have been used for many years in the construction
industry as a tool in the comparison of costs and tender prices
between different periods of time. They are essential when
projecting cash flow forecasts on contracts which may over run
or for planning medium to long term projects where trends from
historical cost data can be identified and projected forwards.

There are several indices available and the most appropriate
must be selected to suit the project under consideration. The main
set of indices in the industry is produced by the Building Cost
Information Services (BCIS) which is published by the RICS.
Other sources include the NEDO indices for the variation of price
clauses prepared by the PSA and published by HMSO.

The Civil Engineering industry under the ICE Conditions of
Contract use a price adjustment formula (sometimes referred to
as the Baxter formula named after the originator) on projects with
a long construction period. This is based on fewer headings than
the Building Cost indices and covers 11 categories.

Construction indices usually present data under two headings;
construction costs and tender prices. Construction costs represent
the amounts paid by contractors for labour, materials, plant and
other costs he must pay to run his business. Tender prices are the
sums accepted by clients for the erection of buildings and include
overheads and profit.

Tender prices increase when there is a large volume of work
available and decrease during lean periods. In difficult times
contractors may submit tenders at cost without any addition for
profit or even overheads to keep the work force together and the
plant occupied.

Tender prices indices are compiled from accepted tenders based on statistics produced in quarterly periods and consolidated into annual figures. The table below lists the construction costs and tender prices between 1976 and 2000 - the figures for 1999 and 2000 are estimated.

Year	Construction costs	Tender prices
1976	100	100
1977	114	107
1978	123	124
1979	142	154
1980	170	190
1981	190	194
1982	210	192
1983	223	199
1984	236	210
1985	248	218
1986	262	229
1987	276	258
1988	293	309
1989	314	340
1990	337	309
1991	355	262
1992	365	243
1993	372	235
1994	382	252
1995	401	265
1996	411	267
1997	421	283
1998	439	313
1999	456	332
2000	474	348

The PSA produce Schedules of Rates which are used for Measured Term Contracts and are updated each month by percentage adjustments An example is set out below to compare the construction costs between 1987 and 1998.

Tender Index for 1987   276

Tender Index for 1998   439

$$\frac{439 - 276}{245} \times 100 = 66.5\%$$

If the construction costs in 1987 were £850,000, the 1998 equivalent figures would be £850,000 × 165.5% - £1,406,750. It may be necessary to take individual elemental costs into account at this stage so that adjustments can be made to ensure that the construction costs of the two buildings are comparable.

REGIONAL VARIATIONS

The cost of construction varies in different parts of the country and this must be taken into consideration in the preparation of first stage estimates. The general level of costs in this book is set at 100.

The following adjustments should be made for each region. These adjustments should be applied to project costs, not individual rates.

England
East Anglia	83
East Midlands	80
Inner London	104
North	81
North West	84
Outer London	98
South East	92
South West	86
Yorkshire and Humberside	81
Northern Ireland	65
Scotland	82
Wales	80

6

Property insurance

When preparing first stage estimates, it is important that the fullest possible financial picture is presented to the client to enable him to make a decision on whether to proceed with the project or not.

This is why the ongoing costs of maintaining buildings must be assessed in a life cycle costing exercise so that they can be added to the construction costs. Similarly, a calculation must be done for insurance purposes to assess the rebuilding costs if the building was damaged by fire or some other cause.

The easiest way to achieve this is by using the square metre prices in Chapter 1 as a base. The appropriate rate should be multiplied by the area of the building to be insured. The resultant figure must then be adjusted by both the indices and regional variation factors in Chapter 5.

The insurance cover must also include for the demolition of the damaged building (not just clearing away debris but grubbing existing foundations and basements) and professional fees to plan and supervise the work of reconstruction. Here is an example of how the calculation should be made.

EXAMPLE

Magistrates' Court in the Yorkshire and Humberside region constructed in 1965 – original cost unknown.

	£
Present day cost: (Chapter 1) 1450 m2 @ £1120	1,624,000
Demolition of old building including site clearance (Chapter 4)	100,000
Carried forward	1,724,000

		£
Brought forward		1,724,000

Regional variation (Chapter 5)
Yorkshire and Humberside
$\times$ 81%   1,396,440

Inflation during the demolition
and rebuilding period based on
a 18 month construction period
and assuming an initial 12 month
pre-building period for the
preparation of contract documents

say 4% $\times$ 2.5 years   <u>139,644</u>   1,536,084

Fees (Chapter 7)

Professional fees say 12%	184,330	
Planning and building		
regulations say	<u>25,000</u>	<u>209,330</u>

		1,745,414
VAT @ 17.5%		<u>305,447</u>
		<u>2,050,861</u>

The building should be insured for £2,100,000.

7

Professional fees

The use of fee scales used to be mandatory but now clients are able to negotiate the level of fees with their professional advisers. In the preparation of first stage estimates, the cost of fees must be included on either an all-in fee basis for the professional team or separate fees for each discipline. This chapter contains information to apply either method.

An allowance should be made for expenses incurred in connection with the contract and VAT on professional fees should also be included. The following categories of fees are included:

- Architects
- Quantity Surveyors
- Consulting Engineers
- Landscape Architects
- Planning
- Building Regulations

ARCHITECTS' FEES

In the following tables, the fees are expressed in a range because of the varying complexities of different projects. There are five classes of work and some examples have been set out below.

Class 1   Sheds and barns
Class 2   Factories, halls and hostels
Class 3   Banks, libraries and schools
Class 4   Restaurants, museums and hospitals
Class 5   Laboratories, theatres and courts

Value of contract	Fee as percentage of contract value
£	%

New work

50,000	7.20 - 9.75
100,000	6.50 - 8.80
150,000	6.20 - 8.45
250,000	5.85 - 8.00
750,000	5.30 - 7.30
1,000,000	5.25 - 7.25
2,000,000	5.15 - 7.05
5,000,000	5.05 - 7.00

Alteration work

50,000	10.55 - 14.75
100,000	9.75 - 13.15
150,000	9.30 - 12.60
250,000	8.85 - 12.00
750,000	8.20 - 11.15
1,000,000	8.10 - 11.00
2,000,000	7.80 - 10.70
5,000,000	7.75 - 10.50

QUANTITY SURVEYORS' FEES

The following examples of fees are set out in Professional Charges for Quantity Surveying Services obtainable from Surveyors Publications, Surveyors' Court, Westwood Way, Coventry CV4 8JE, tel. 01203 694757.

Scales 36 and 37 – Building work

There are three basic categories of works:

Category A	complex with little repetition
Category B	less complex with some repetition
Category C	simple.

For each category there are two types of fees:

Scale 36   inclusive scale for complete services

Scale 37   itemised scale divided into pre- and post-contract services.

Here are some examples of the percentages and fees for varying sized projects.

Scale 36 Inclusive services

Value of Work £	Category A £	%	Category B £	%	Category C £	%
150,000	9,380	6.25	9,060	6.04	7,650	5.10
250,000	14,380	5.75	13,760	5.50	11,750	4.70
350,000	19,030	5.44	18,060	5.16	15,450	4.41
450,000	23,330	5.18	21,960	4.88	18,750	4.17
750,000	34,880	4.65	32,010	4.27	27,450	3.66
1,250,000	51,880	4.15	46,010	3.68	39,950	3.20
2,500,000	90,380	3.62	79.010	3.16	68,200	2.73
4,000,000	133,380	3.33	116,010	2.90	99,200	2.48

Scale 37 Pre-contract services

Value of Work £	Category A £	%	Category B £	%	Category C £	%
150,000	4,730	3.15	4,410	2.94	3,930	2.62
250,000	7,030	2.81	6,410	2.56	5,730	2.29
350,000	9,080	2.59	8,160	2.33	7,230	2.27
450,000	10,880	2.42	9,660	2.15	8,430	4.17
750,000	15,830	2.11	13,560	1.81	11,580	1.54
1,250,000	23,330	1.87	19,060	1.52	16,080	1.29
2,500,000	39,080	1.56	31,810	1.27	26,330	1.05
4,000,000	56,380	1.40	45,810	1.15	37,330	0.93

Scale 37 Post-contract services (Alternative 1)

Value of Work £	Category A £	%	Category B £	%	Category C £	%
150,000	3,150	2.10	3,150	2.10	2,520	1.68
250,000	4,850	1.94	4,850	1.94	4,020	1.61
350,000	6,500	1.86	6,450	1.84	5,470	1.56
450,000	8,100	1.80	7,950	1.77	6,870	1.53
750,000	12,450	1.66	11,850	1.58	10,620	1.42
1,250,000	18,950	1.52	17,350	1.52	16,120	1.29
2,500,000	34,200	1.37	30,100	1.20	27,870	1.11
4,000,000	51,200	1.28	44,100	1.10	40,370	1.01

For negotiating and agreeing prices with a contractor

Value of Work £	Fee £	%
150,000	750	0.50
250,000	1,050	0.42
350,000	1,350	0.39
450,000	1,650	0.37
750,000	2,400	0.32
1,250,000	3,350	0.27
2,500,000	4,600	0.18
4,000,000	6,100	0.15

Scale 38 – Civil Engineering work

Category I   runways, roads, railways and earthworks and dredging and monolithic walls.

Category II   piled quay walls, suspended jetties, bridges, sewers, storage and treatment tanks, turbine halls, reactor blocks.

Pre-contract services

Value of Work	Category I Fee		Category II Fee	
£	£	%	£	%
500,000	1,960	0.65	3,650	0.73
1,500,000	5,040	0.34	8,850	0.50
2,500,000	7,540	0.30	12,850	0.51
5,000,000	12,790	0.26	21,850	0.44
7,000,000	16,790	0.24	28,850	0.41
12,000,000	25,790	0.21	45,350	0.38
15,000,000	30,290	0.20	54,350	0.36
25,000,000	44,790	0.18	83,350	0.33

Post-contract services

Value of Work	Category I Fee		Category II Fee	
£	£	%	£	%
500,000	5,950	1.19	10,750	2.15
750,000	8,250	1.10	15,000	2.00
1,500,000	14,250	0.95	26,250	1.75
2,500,000	20,750	0.83	38,250	1.53
5,000,000	34,000	0.68	65,250	1.31
7,000,000	44,000	0.63	86,250	1.23
12,000,000	68,000	0.57	135,750	1.13
15,000,000	81,500	0.54	162,750	1.09
25,000,000	125,000	0.50	249,750	1.00

CONSULTING ENGINEERS' FEES

Engineers' fees are calculated by dividing the Cost of the Works by the Output Price Index and the resultant figure is then applied to a graph to show the percentage to be set to the final cost of the works. Full details can be obtained from The Association of Consulting Engineers, Alliance House, 12 Caxton Street, London SW1H OQZ , tel. 0171 222 6557.

LANDSCAPE CONSULTANTS' FEES

The fees for work over the value of £10,000 are calculated in two parts. Part 1 is assessed from a graph which indicates the fee percentage from 6% to 14% according to the value of the contract. Part 2 is a coefficient ranging from 1.0 where the consultant has overall responsibility to his client for a job with a normal balance of hard and soft works.

This may increase to 1.2 when the soft works element exceeds 50% of the landscape contract or for private garden contracts. The coefficient may be decreased to 0.8 for other types of jobs such as golf courses and road landscaping. Here is an example.

Assume a project has a contract value of £100,000 including both hard and soft works in a new business park. The fee graph shows that the percentage norm is 7.5%. The coefficient is 1.2 because the soft work element exceeds 50%. The job coefficient is 1.0 so the compounded coefficient is 1.0 × 1.2. The total percentage fee will be 7.5% × 1.2 = 9.0%.

The scale of fees allows for other methods of remuneration such as lump sum fees, using a ceiling figure in conjunction with a time basis or having a retainer which can be reviewed after a period and paid according to the value of the actual work carried out .

Similarly, when only occasional work is required this can be charged on a time basis. Site surveys would normally be paid on a lump sum basis of estimated time involved.

PROFESSIONAL TEAM ALL-IN FEES

Assessing the fees for a professional team working on a development project can be complicated due to the different methods of fee calculation adopted by the various bodies. Worked examples of these are shown in this chapter relating to a range of contract values.

Sometimes the client will prefer to deal with only one discipline and a lead professional will be appointed. This firm will negotiate an overall fee for the whole team and the table below shows the effect of this arrangement. The figures have been rounded off to the nearest £1,000.

Project cost £	8% £000	9% £000	10% £000	11% £000	12% £000	13% £000	14% £000	15% £000
200	16	18	20	22	24	26	28	30
300	24	27	30	33	36	39	42	45
400	32	36	40	44	48	52	56	60
500	40	45	50	55	60	65	70	75
600	48	54	60	66	72	78	84	90
700	56	63	70	77	84	91	98	105
800	64	72	80	88	96	104	112	120
900	72	81	90	99	108	117	126	135
1,000	80	90	100	110	120	130	140	150
1,200	96	108	120	132	144	156	168	180
1,400	112	126	140	154	168	182	196	210
1,600	128	144	160	176	192	208	224	240
1,800	144	162	180	198	216	234	252	270
2,000	160	180	200	220	240	260	280	300
2,250	180	202	225	247	270	292	315	337
2,500	200	225	250	275	300	325	350	375
2,750	220	247	275	302	330	357	385	412
3,000	240	270	300	330	360	390	420	450
3,250	260	292	325	357	390	422	455	487
3,500	280	315	350	385	420	455	490	525
3,750	300	337	375	412	450	487	525	562
4,000	320	360	400	440	480	520	560	600
5,500	440	495	550	605	660	715	770	825
6,000	480	540	600	660	720	780	840	900
7,000	560	630	700	770	840	910	980	1050
8,000	640	720	800	880	960	1040	1120	1200
9,000	720	810	900	990	1080	1170	1260	1350
10,000	800	900	1000	1100	1200	1300	1400	1500

PLANNING PERMISSION FEES

These can be quite complex and a summary of some of the main fees is set out.

Dwelling houses	£190 per 0.1 hectare for outline planning (maximum £4,750) and £190 per house (maximum £9,500).

Alterations and extensions to houses	Single house £95. Two or more houses £190.
Change of use of building for houses	£190 per house (maximum £9,500).

BUILDING REGULATION FEES

Local authorities are now able to set their own scale of fees for work in connection with building regulations. There are three main charges involved:

> *Plan Charge* or *Full Plan* to be paid when plans are submitted

> *Building Notice* to be paid when the Notice is submitted

> *Inspection Fee* to be paid after first inspection.

Here are some examples of these charges which include VAT.

Small domestic work and alterations

Work	Plan Charge £	Building Notice £	Inspection Fee £
Extension less than 10m2	120	nil	120
Extension 10 - 40m2	80	180	250
Extension 10 - 40m2	100	275	350
Detached garages	20	70	90

8

Life cycle costing

Life cycle costing is a system for budgeting and controlling the costs of the design, development and property management of a building. It should focus on the building owner's policy requirement for the building.

The life cycle cost of a building is the total cost commitment to that building and is the sum of the initial capital costs and future running costs. These costs are incurred at different times so they need to be discounted back to the present to allow them to be compared to the initial capital costs. These are standard techniques and tables are published for discounting future costs and expressing them as a single sum of money (Net Present Value or NPV) or as an annual flow of money (Annual Equivalent).

A range of terms are used in life cycle costing and the meanings of the most common terms are defined below.

Life cycle cost planning	The systematic and objective quantification of the initial and life time costs of all of a building's elements at the design stage to produce a scheme which satisfies the client's objectives.
Cost-in-use	The technique of converting the life time cost consequences of a particular design decision into a single consistent measure of cost which can be used for comparison purposes. The measure may be a single sum of money (NPV), a flow of money per time period (Annual Equivalent) or it may be expressed in terms of scarce resources whose use is to be minimised such as energy or manpower.

Life cycle cost analysis	The systematic collection and analysis of the running costs and performance characteristics of a building in use. The analysis should reflect the degree to which the costs incurred are justifiable in terms of the performance delivered.
	The purpose of the analysis is to provide appropriately structured and processed data which can be used to prepare life cycle cost plans for new buildings or the life cycle cost management of an existing building.
Life cycle cost management	The process for developing and implementing a maintenance policy which recognises that the building owner's or occupier's interest may be served best by setting maintenance decisions into his broader financial context and in a time frame which goes beyond the immediate maintenance horizon. The purpose of life cycle cost management is to establish maintenance management as a creative activity rather than one which is routine or reactive.

The inherent weakness of life cycle cost studies is that the formative influences on a building's life cycle characteristics are the decisions taken at the early stages when there is little or no project specific information available which can be used to test the validity of those early decisions.

Furthermore, the appraisal of design options for life cycle costing purposes must assume that cost implications can be reliably predicted for many years into the future. This is a somewhat fragile assumption given the long life of buildings and that the costs themselves are the result of a complex interplay between performance, use behaviour and changes in the economic environment.

The main emphasis for the developer or owner of the buildings should be to concentrate on generating an outline brief and management policy in life cycle terms. Concentrating on the outline brief has the significant benefit of defining the objectives for the completed development at a stage when they are still capable of affecting design decisions.

The attention paid to formulating a management policy recognises the fact that costs over the life of a building become less predictable the further they are removed from the present. The owner's interests may be served by adopting a system which can deal with these future costs when they arise in a manner which is consistent with the project's objectives, which will have developed from the initial definition.

The intermediate stages between the outline brief at the commencement of design and the management policy for the building in use, should be used to translate the original requirements into a physical structure which is sufficiently robust to allow subsequent management decisions to moderate its performance in response to evolving objectives and environment conditions.

The benefits of adopting a life cycle approach comes from the recognition that buildings are a long term investment. Assessing buildings' value from the initial capital costs is inconsistent with their long term nature and so attention should also be given to the future costs of buildings. Their design should result from decisions which recognises the objectives for the building over its whole life cycle.

Here is a checklist for life cycle costing at the outline brief stage when preparing a Life Cycle Cost Plan. It is essential that the following project's objectives are established:

client's interest, e.g. owner occupation /develop and lease/ develop and sell

client's time horizon
    short term – develop
    medium term – develop and own for a limited period
    long term – develop and own long term

time horizon over which client's objectives will remain stable

client's specific space requirements if for owner occupation nature of tenants, if for leasing, e.g. known tenants with known preferences – speculative tenant groups with open requirements

nature of lease and services charges. If tenanted, type and duration of lease, rent and service charge mix, division of responsibility between owner and tenants and basis of apportionment between tenants

importance of the project realising a capital asset and over what time scale this realisation should take place

if there is a proposed purchaser and, what are his preferred requirements

where the project fits into the client's financial strategy

the proposed method of finance, e.g. development finance, finance during holding period and long-term finance

impact of project on client's cash flow, e.g. during development, during ownership and during redevelopment, refurbishment and disposal

effects on client's capital structure, e.g. gearing and other options, effects on future financing options and effects on market capitalisation both current and future

where the project fits into the client's main business strategy

what resources the client can apply to develop the project, actively manage the project (who and how) and manage the completed development.

The time-related variables:

the life cycles, e.g. physical life of structure and key elements, economic life related to underlying land values, functional life, social life and leasehold cycles

the coincidence of the client's requirements with the above life cycles

time-related operational constraints, e.g. the criticality of the completion date

the time-related marketing constraints

the time-related financial constraints, e.g. the running out of a line of credit

the time-related contractual constraints

the time-related statutory constraints, e.g. changes in taxation or building regulations

time-related political constraints.

The finance-related variables:

the client's investment criteria, e.g. payback – DCF – ROCE

whether the criteria measure is what the client is trying to achieve

the financial objectives, e.g. maximise potential for rental income, maximise potential for cash generation, minimise running costs for owner occupier and secure long term capital growth

whether the funding arrangements mean that the project must achieve a particular asset value within a specific period of time

the client's capital and revenue constraints and how they impact on the viability of the project, e.g. cost yardsticks, financial or operational criteria and allowances for enhancement or avoidance

whether the residual value of the project has a significant impact on its viability. If so, is this at the termination of the project's life, at significant intervals or annually for valuation purposes

the relationship between the value of the project and the underlying land value

whether the ownership strategy envisages funds becoming available during the life of the project to counteract the effects of value depreciation relative to newer buildings coming on to the market

to what extent there is a trade-off between capital and revenue

the client's attitude to risk both during development and ownership

the importance of the client's tax position regarding the timing of cash flows, the capital cost/running cost relationship Capital Gains Tax or relief therefrom

whether the tax status of the prospective tenant or purchaser impacts significantly in the client's objectives

what are the tax implications surrounding the project, e.g. capital allowances and rates

whether the location of the project has fiscal implications, e.g. grants, tax allowances and rates relief.

The design-related variables:

the extent of finish for the completed development, e.g. shell and core, ready to decorate or ready to occupy

the quality of finish for the completed development, e.g. external envelope, internal furnishings and fixtures, public and common areas and general accommodation the degree of structural flexibility required to cater for probable or possible future developments

the degree of superficial flexibility required to cater for changes in the image or function of the building

the importance of cost and time trade-offs to the maintenance and refurbishment of the building in use, e.g. will disruption costs and rental losses be high

whether the local economy can provide for the type of construction, the quality of fixture and fittings and the level of maintenance envisaged.

The external variables:

a weighted discount rate. e.g. short-term, long-term, relative to the source of finance and relative to opportunity costs

inflation rates for project costs, operating costs, project revenues and tax position of client or occupier

commercial risk and relate to choice of contract, maintenance contract and best and worst scenarios of key cost elements

political risks, e.g. potential changes in direct taxation and tax allowances, potential changes in indirect taxation, potential changes in local taxation and susceptibility of financial arrangements to political change.

Once the building has been occupied, information should be generated defining its true performance in use. At this stage it should be possible to establish a programme of maintenance, repair and refurbishment which accurately reflects the interplay between the client's objectives and the performance of the building, i.e. life cycle management. A worked example of a life cycle cost plan is set out below.

Background information

Job title:	New warehouse
Location:	North West England
Client:	Private development
Date:	January 2000
Real discount rate:	9% (cost of capital = 10%; inflation @ 1.5% discount rate = 1.10/1.015%
Life cycle:	20 years
Job details:	Steel portal framed single storey building, external works and drainage
Areas:	60,000 sq. ft

Life Cycle Cost Plan

Capital cost   £

Building; 60,000 sq. ft. @ £35	=	2,100,000
External works (approx. quants)	=	300,000
Contingencies @ 25% (client; design; contract)	=	500,000
Design fee @ 7%	=	200,000
		3,100,000

Operating costs (annual)   £

General and water rates p.a.	=	100,000
Present value (20 years)	=	910,000
Heating and lighting p.a.	=	150,000
Present value	=	1,370,000
Maintenance and insurance p.a.	=	35,000
Present value	=	320,000
		2,885,000

Intermittent repair and replacement costs	**£**
Redecorate every 3 years @ £10,000, present value (every 3 years at 9%)	45,000
Re-carpet every 5 years @ £15,000, present value (every 5 years at 9%)	40,000
Replace furniture every 7 years @ £50,000, present value (every 7 years at 9%)	90,000
Lighting and power refurbishment every 10 years @ £25,000, present value (every 10 years at 9%)	45,000
	220,000

Summary of costs	**£**
Capital cost	3,100,000
Running and operating costs	2,885,000
Repair and replacement costs	220,000
Total Present Value	**6,205,000**

9

Useful addresses

Architects Registration Council of the United Kingdom
73 Hallam Street,
London W1 N6EE
(0207-580 5861)

Association of Consulting Engineers
Alliance House,
12 Caxton Street,
London SW1 H1QL
(0207-222 6557)

Brick Development Association
Woodside House,
Winkfield, Windsor,
Berkshire SL4 2DP
(01344 885651)

British Board of Agrement
PO Box 195, Bucknall's Lane,
Garston, Watford,
Herts WD2 7NG
(01923 670844)

British Computer Society
1 Sansord Street,
Swindon, Wilts SN1 1HJ
(01793 417417)

British Property Federation
35 Catherine Place,
London SW1E 6DY
(0207-828 0111)

British Standards Institution
389 Chiswick High Street,
London W4 4AL
(0207-629 9000)

British Steel Plc
9 Albert Embankment,
London SE1 7SN
(0207-735 7654)

British Woodworking Federation
82 New Cavendish Street,
London W1M 8AD
(0207-872 8210)

Builders' Merchants Federation
15 Soho Square,
London W1V 5FB
(0207-439 1753)

Building Centre
26 Store Street,
London WC1E 7BT
(0207-637 1022/8361

Building Employers Confederation
18 Duchess Mews,
London W1
(0207-636 3891)

**Building Research Establishment
(Department of the Environment),**
Bucknall's Lane, Garston,
Watford WD2 7JR
(01923 664000)

Central Office of Information
Hercules Road,
London SE1 7DU
(0207-928 2345

Chartered Institute of Arbitrators
24 Angle Gate
London EC1
(0207 837 4483)

Chartered Institute of Building
Englemere, King's Ride,
Ascot, Berkshire SL5 8BJ
(01344 630700)

Confederation of British Industry
Centre Point,
103 New Oxford Street,
London WC1
(0207-379 7400)

Electrical Contractors Association
ESCA House,
34 Palace Court,
London W2
(0207-229 1266)

Federation of Building Sub-Contractors
82 New Cavendish Street,
London W1M 8AD
(0207-580 5588)

Federation of Master Builders
14 Great James Street,
London WC1N 2DP
(0207-242 7583)

Glass Manufacturers' Federation
19 Portland Place,
London W1N 4BH
(0207-580 6952)

Heating and Ventilation Contractors Association
ESCA House,
34 Palace Court,
London W2 4JG
(0207-229 2488)

Housing Corporation
149 Tottenham Court Road,
London W1P OBN
(0207-393 2000)

Institute of Mechanical Engineers
1 Birdcage Walk,
London SW1H 9JJ
(0207-222 7899)

Institute of Plumbing
64 Station Lane,
Hornchurch,
Essex RN12 6NB
(01708 472791)

Institution of Civil Engineers
1-7 Great George Street,
London SW1P 3AA
(0207-222 7722)

Institution of Civil Engineering Surveyors
Dominion House,
Sibson Road,
Sale,
Cheshire M33 7PP
(0161-972 3100)

Institution of Electrical Engineers
2 Savoy Place,
London WC2R OBL
(0207-240 1871)

Institution of Structural Engineers
11 Upper Belgrave Street,
London SW1X 8BH
(0207-235 4535/6841)

Iron and Steel Trades Confederation
Swinton House,
324 Grays Inn Road,
London WC1X 8DD
(0207-837 6691)

**Joint Industry Board for the Electrical
Contracting Industry**
Kingswood House,
47/51 Sidcup Hill,
Sidcup,
Kent DA14 6HJ
(0208 302 0031)

National Association of Local Councils
109 Great Russell Street,
London WC1B 3LD
(0207-637 1865)

**National Association of Plumbing, Heating
and Mechanical Services Contractors**
6 Gate Street,
London WC2A 3HX
(0207-405 2678)

National Association of Scaffolding Contractors
18 Mansfield Street,
London W1M 9FG
(0207-580 558)

National Association of Shopfitters
NAS House, 411 Limpsfield Road,
The Green, Warlingham,
Surrey CR3 9HA
(01883 624961)

National Building Specification
Mansion House Chambers,
The Close,
Newcastle-upon-Tyne NE1 3OE
(0191-232 9594)

National Computing Centre
Oxford Road,
Manchester M1 7ED
(0161-242 2100)

National Council of Building Material Producers
26 Store Street,
London WC1E 7BT
(0207-323 3770)

National Federation of Housing Associations
175 Grays Inn Road,
London WC1X 8UP
(0207-278 6571)

National Federation of Demolition Contractors
1A New Road,
The Causeway,
Staines, Middlesex TW18 3DH
(0207-404 4020)

**National Federation of Painting and
Decorating Contractors**
18 Mansfield Street,
London W1M 9FG
(0207-580 5588)

National Federation of Plastering Contractors
82 New Cavendish Street,
London W1M 8AD
(0207-580 5588)

National Federation of Roofing Contractors
24 Weymouth Street,
London W1N 4LX
(0207-436 0387)

**National Joint Council for Felt Roofing
Contracting Industry**
Fields House,
Gower Road,
Haywards Heath,
West Sussex RH16 4PL
(01444 440027)

Royal Institute of British Architects
66 Portland Place,
London W1N 4AD
(0207-580 5533)

Royal Institute of Chartered Surveyors
12 Great George Street,
London SW1Y 5AG
(0207-222 7000)

The Brick Development Association
Woodside House,
Winkfield,
Windsor SL4 2DX
(013447 885651)

Town and Country Planning Association
17 Carlton House Terrace,
London SW1Y 5AS
(01-930 8903/5)

Water Authorities Association
1 Queen Ann's Gate,
London SW1H 9BT
(0207-957 4567)

Welsh Development Agency
Treforest Industrial Estate,
Pontypridd,
Glamorgan CS37 5UT
(01345 775577)

Welsh Office
Cathays Park,
Cardiff CF1 3NQ
(01222 825111)

Zinc Development Association
42 Weymouth Street,
London W1N 3LQ
(0207-499 6636)

Estimating data

When preparing first stage estimates it is often necessary to take a broad view because of lack of time or information. The data in this chapter is intended to help in this process.

LABOUR

The wage awards taking effect from 29 June 1999 were £235.95 for craftsmen and £177.45 for labourers. A model calculation to produce an hourly rate is set out below.

		£
Flat time	2028 hours @ £6.20	12,573.36
NIC Employer's contribution	7%	880.15
Severance pay and statutory costs	2%	251.47
CITB levy	0.25%	31.43
Non-productive overtime	39 hours @ £6.20	241.80
Public holidays	39 hours @ £6,20	241.80
Sickness, injury holidays and death benefit	47 weeks @ £20.95	984.65
Employers' liability	2%	251.47
		15,456.13
divided by 1864 hours		8.29

For estimating purposes, craftsmen should be costed at, say £8.50 per hour and labourers at £6.50. Hourly rates for other trades have been set out below.

Plumber	£10.50
Mechanical engineer	£11.50
Electrician	£12.00

MATERIALS

Weights of materials

Aggregate, coarse	1500kg/m3
Aluminium, 0.80mm thick	2kg/m2
Ashes	800kg/m3
Ballast	600kg/m3
Blockboard	
standard	1000kg/m2
tempered	1060kg/m2
Blocks, lightweight aggregate	
75mm thick	60kg/m2
100mm thick	80kg/m2
140mm thick	112kg/m2
Blocks, natural aggregate	
75mm thick	160kg/m2
100mm thick	215kg/m2
140mm thick	300kg/m2
Bricks, Fletton	1820kg/m2
engineering	2250kg/m2
concrete	1850kg/m2
Brickwork, 112.5mm	220kg/m2
215 mm	465kg/m2
327.5mm	710kg/m2
Carlite browning, 11mm thick	7.80kg/m2
Carlite bonding coat	
8mm thick	7.10kg/m2
11mm thick	9.80kg/m2
Cement	1440kg/m3
Chalk	2240kg/m3

Clay	1800kg/m3
Concrete	2450kg/m3
Copper, 0.55mm thick	5kg/m2
Glass, float	
3mm	7kg/m2
4mm	10kg/m2
5mm	13kg/m2
6mm	15kg/m2
10mm	25kg/m2
12mm	30kg/m2
15mm	38kg/m2
19mm	48kg/m2
25mm	64kg/m2
Glass, clear sheet	
3mm	7kg/m2
4mm	10kg/m2
5mm	13kg/m2
6mm	15kg/m2
Glass, patterned	
3mm	6kg/m2
4mm	8kg/m2
5mm	10kg/m2
6mm	12kg/m2
10mm	22kg/m2
Flint	2550kg/m3
Gravel	1750kg/m3
Hardcore	1900kg/m3
Hoggin	1750kg/m3
Lead sheeting, 2.24mm thick	25kg/m2
Lime, ground	750kg/m3
Pipes, PVC	
80mm	1.20kg/m
110mm	1.60kg/m
160mm	3.00kg/m
200mm	4.60kg/m
250mm	7.20kg/m
Pipes, vitrified clay	
100mm	15.63kg/m
150mm	37.04kg/m
225mm	95.24kg/m

300mm	196.08kg/m
400mm	357.14kg/m
450mm	500.00kg/m
500mm	555.60kg/m
Sand	1600kg/m3
Steel bars, flat	
25 × 9.53	1.910kg/m
38 × 9.53	2.840kg/m
50 × 12.70	5.060kg/m
50 × 19.00	7.590kg/m
Steel bars, round	
6mm	0.283kg/m
8mm	0.503kg/m
10mm	0.784kg/m
12mm	0.130kg/m
16mm	2.010kg/m
20mm	3.139kg/m
25mm	4.905kg/m
32mm	8.035kg/m
40mm	12.554kg/m
50mm	19.667kg/m
Steel bars, square	
6mm	0.222kg/m
8mm	0.395kg/m
10mm	0.616kg/m
12mm	0.888kg/m
16mm	1.579kg/m
20mm	2.466kg/m
25mm	3.854kg/m
32mm	6.313kg/m
40mm	12.544kg/m
50mm	15.413kg/m
Stone, crushed	1350kg/m3
natural	2400kg/m3
reconstructed	2250kg/m3
Terrazzo, 25mm thick	45.50kg/m2
Timber	
ash	800kg/m3
Baltic spruce	480kg/m3
beech	816kg/m3

birch	720kg/m3
box	961kg/m3
cedar	480kg/m3
ebony	1217kg/m3
elm	624kg/m3
greenheart	961kg/m3
jarrah	816kg/m3
maple	752kg/m3
oak, American	720kg/m3
oak, English	848kg/m3
pine, pitchpine	800kg/m3
pine, red deal	576kg/m3
pine, yellow deal	528kg/m3
sycamore	530kg/m3
teak, African	961kg/m3
teak, Indian	656kg/m3
walnut	496kg/m3
Water	950kg/m3
Woodblock flooring	
softwood	12.70/kgm2
hardwood	17.60kg/m2
Wood chipboard	
standard grade	700kg/m2
flooring grade	750kg/m2
Zinc 0.65mm thick	4kg/m2

EXCAVATION

Shrinkage of deposited materials

Clay	−10.0%
Gravel	−7.5%
Sandy soil	−12.5%

Bulking of excavated material

Clay	+ 40%
Gravel	+ 25%
Sand	+ 20%

Angle of repose

Angle of repose	Type	Angle °
Earth	loose, dry	36-40
	loose, moist	45
	loose, wet	30
	consolidated, dry	42
	consolidated, moist	38
Loam	loose, dry	40-45
	loose, wet	20-25
Gravel	dry	35-45
	wet	25-30
Sand	loose, dry	35-40
	compact	30-35
	wet	25
Clay	loose, wet	20-25
	consolidated, moist	70

CONCRETE WORK

Concrete mixes (per m3)

Mix	Cement t	Sand m3	Aggregate m3	Water litres
1:1:2	0.50	0.45	0.70	208
1:1.5:3	0.37	0.50	0.80	185
1:2:4	0.30	0.54	0.85	175
1:2.5:5	0.25	0.55	0.85	166
1:3:6	0.22	0.55	0.85	160

Grade

Grade	Cement t	Sand m3	Aggregate m3	Water litres
20/20	0.32	0.62	1.20	170
25/20	0.35	0.60	1.17	180
30/20	0.80	0.59	1.11	200
7/40 all-in	0.18	-	1.95	150
20/20 all-in	0.32	-	1.85	170
25/20 all-in	0.36	-	1.75	180

Steel bar reinforcement

Diameter mm	Nominal weight kg/m	Length m/tonne	Sectional area mm2
6	0.222	4505	28.30
8	0.395	2532	50.30
10	0.616	1623	78.50
12	0.888	1126	113.10
16	1.579	633	201.10
20	2.466	406	314.20
25	3.854	259	490.90
32	6.313	158	804.20
40	9.864	101	1256.60
50	15.413	65	1963.50

BRICKWORK

Bricks per m2
(brick size 215 ×102.5 × 65mm)   **nr**

	nr
Half brick wall	
stretcher bond	59
English bond	89
English garden wall bond	74
Flemish bond	79
One brick wall	
English bond	118
Flemish bond	118
One and a half brick wall	
English bond	178
Flemish bond	178
Two brick wall	
English bond	238
Flemish bond	238

Metric modular bricks
per square metre

	nr
200 × 100 × 75mm	67
90mm thick	133
190mm thick	200
200 × 100 × 100mm	
90mm thick	50
190mm thick	100
290mm thick	150
300 × 100 × 75mm	
90mm thick	44
300 × 100 × 100mm	
90mm thick	33

Blocks per m2
(block size 414 × 215mm)

	nr
60mm thick	9.9
75mm thick	9.9
100mm thick	9.9
140mm thick	9.9
190mm thick	9.9
215mm thick	9.9

Mortar per m2

Brick size	Wirecut m3	One frog m3
215 × 102.5 × 65mm		
half brick wall	0.017	0.024
one brick wall	0.045	0.059
one and a half brick wall	0.072	0.093
two brick wall	0.101	0.128

Brick size	Solid m3	Perforated m3
200 × 100 × 75mm		
90mm thick	0.016	0.019
190mm thick	0.042	0.048
290mm thick	0.068	0.078
200 × 100 × 100mm		
90mm thick	0.013	0.016
190mm thick	0.036	0.041
290mm thick	0.059	0.067
300 × 100 × 75mm		
90mm thick	0.015	0.018
300 × 100 × 100mm		
90mm thick		0.015

Block size	m3
440 × 215mm	
60mm	0.004
75mm	0.005
100mm	0.006
140mm	0.007
190mm	0.008
215mm	0.009
Random rubble walling	
300mm thick	0.120
450mm thick	0.160
550mm thick	0.200

**Length of pointing per m2
(one face only)**

English bond	19.1m
English garden wall bond	18.1m
Flemish bond	18.4m
Flemish garden wall bond	17.7m

ROOFING

	Lap mm	Gauge mm	nr/m2	Battens m/m2
Clay/concrete tiles				
267 × 165mm	65	100	60.00	10.00
	65	98	64.00	10.50
	65	90	68.00	11.30
387 × 230mm	75	300	16.00	3.20
	100	280	17.40	3.50
420 × 330mm	75	340	10.00	2.90
	100	320	10.74	3.10
Fibre slates				
500 × 250mm	90	205	19.50	4.85
	80	210	19.10	4.76
	70	215	18.60	4.65
600 × 300mm	105	250	13.60	4.04
	100	250	13.40	4.00
	90	255	13.10	3.92
	80	260	12.90	3.85
	70	263	12.70	3.77
400 × 200mm	70	165	30.00	6.06
	75	162	30.90	6.17
	90	155	32.30	6.45
500 × 250mm	70	215	18.60	4.65
	75	212	18.90	4.72
	90	205	19.50	4.88
	100	200	20.00	5.00
	110	195	20.50	5.13
600 × 300mm	100	250	13.40	4.00
	110	245	13.60	4.08
Natural slates				
405 × 205mm	75	165	29.59	8.70
405 × 255mm	75	165	23.75	6.06
405 × 305mm	75	165	19.00	5.00
460 × 230mm	75	195	23.00	6.00
460 × 255mm	75	195	20.37	5.20
460 × 305mm	75	195	17.00	5.00

	Lap mm	Gauge mm	nr/m2	Battens m/m2
510 × 255mm	75	220	18.02	4.60
510 × 305mm	75	220	15.00	4.00
560 × 280mm	75	240	14.81	4.12
560 × 305mm	75	240	14.00	4.00
610 × 305mm	75	265	12.27	3.74

Reconstructed stone slates

	Lap mm	Gauge mm	nr/m2	Battens m/m2
380 × 2150mm	75	150	16.00	3.20
	100	140	17.40	3.50

Lead work	Code	Colour	Thickness	kg/m2 mm
1.32mm	3	Green	1.32	14.97
1.80mm	4	Blue	1.80	20.41
2.24mm	5	Red	2.24	25.40
2.65mm	6	Black	2.65	30.05
3.15mm	7	White	3.15	35.72
3.55mm	8	Orange	3.55	40.26

CARPENTRY

Lengths of boarding metres per m2

Effective width mm	m/m2
75	13.33
100	10.00
125	8.00
150	6.60
175	5.71
200	5.00

PLASTERING AND TILING

Tiles per m2	nr
150 × 150mm	44.36
200 × 200mm	25.00
300 × 300mm	11.09
500 × 500mm	4.00

Plaster coverage	Thickness mm	m2/1000kg
Carlite browning	11	135-155
Carlite tough coat	11	135-150
Carlite bonding	11	100-115
Carlite hardwall	11	115-130
Thistle dri-coat	11	125-135
Thistle renovating	11	115-125
Thistle universal	13	85-95

GLAZING

Putty per m2	Wood kg/m2	Metal kg/m2
Panes up 0.10m2	3.95	5.08
	4.21	5.42
	4.56	5.86
	4.96	6.38
	5.33	6.86
Panes 0.10-0.50m2	1.62	2.09
	1.72	2.21
	1.87	2.41
	2.03	2.61
	2.18	2.80
Panes 0.50-1.00m2	1.03	1.32
	1.09	1.40
	1.19	1.52
	1.28	1.64
	1.38	1.77

Putty per m2	Wood kg/m2	Metal kg/m2
Panes over 1.00m2	0.73	0.93
	0.74	0.99
	0.77	1.08
	0.91	1.17
	0.97	1.25

WALLPAPERING

Wallpaper roll sizes	Length m	Width m	Area m2
UK	10.00	0.53	5.30
France	11.00	0.57	6.27
USA	7.31	0.46	3.36
Europe	10.65	0.71	7.56

Rolls required

Room perimeter m	Wall height m	Rolls required nr
8	2.5	4
9	2.5	5
10	2.5	5
11	2.5	6
12	2.5	6
13	2.5	7
14	2.5	7
15	2.5	8
16	2.5	8
17	2.5	8
18	2.5	9
19	2.5	10
20	2.5	10
21	2.5	10
22	2.5	11
23	2.5	11

DRAINAGE

Volumes of filling (m3 per linear metre)

Pipe dia. mm	Beds			Bed and haunching	Surround
	50mm m3	100mm m3	150mm m3	m3	m3
100	0.023	0.045	0.068	0.117	0.185
150	0.026	0.053	0.079	0.152	0.231
225	0.030	0.060	0.090	0.195	0.285
300	0.038	0.075	0.113	0.279	0.391

Trench widths

Pipe dia. mm	Not exc. 1.5m deep mm	Exceeding 1.5m deep mm
100	450	600
150	500	650
225	600	750
300	650	800
400	750	900
450	900	1050
600	1000	1300

PLANT

Typical fuel consumption for plant

These figures relate to working in normal conditions. Reduce by 25% for light duties and increase by 50% for heavy duties.

Plant	Engine size kW	Litres/ hour
Compressors up to	20	4.0
	30	6.5
	40	8.2

Plant	Engine size kW	Litres/ hour
	50	9.0
	75	16.0
	100	20.0
	125	25.0
	150	30.0
Concrete mixers up to	5	1.0
	10	2.4
	15	3.8
	20	5.0
Dumpers	5	1.3
	7	2.0
	10	3.0
	15	4.0
	20	4.9
	30	7.0
	50	12.0
Excavators	10	2.5
	20	4.5
	40	9.0
	60	13.0
	80	17.0
Pumps	5	1.1
	8	1.6
	10	2.1
	15	3.2
	20	4.2
	25	5.5
Trenchers	25	5.0
	35	6.5
	50	10.0
	75	14.5

Average plant outputs (m3/hour)

Bucket size (litres)	Soil m3	Sand m3	Heavy clay m3	Soft rock m3
Face shovel				
200	11	12	7	5
300	18	20	12	9
400	24	26	17	13
600	42	45	28	23
Backactor				
200	8	8	6	4
300	12	13	9	7
400	17	18	11	10
600	28	30	19	15
Dragline				
200	11	12	8	5
300	18	20	12	9
400	25	27	16	12
600	42	45	28	21

MEASUREMENT

Metric conversion tables

To convert:	Multiply by:
Length	
Inches to centimetres	2.54
Centimetres to inches	0.3937
Inches to millimetres	25.4
Millimetres to inches	0.03937
Feet to metres	0.3048
Metres to feet	3.2808
Yards to metres	0.9144
Metres to yards	1.0936
Miles to kilometres	1.6093
Kilometres to miles	0.6213

Area

Sq inches to sq centimetres	6.4516
Sq centimetres to sq inches	0.155
Sq feet to sq metres	0.0929
Sq metres to sq feet	10.7639
Sq yards to sq metres	0.8361
Sq metres to sq yards	1.1959
Acres to hectares	0.4046
Hectares to acres	2.471
Sq miles to sq kilometres	0.3861
Sq kilometres to sq miles	0.3861

Volume

Cu inches to cu centimetres	16.387
Cu centimetres to cu inches	0.06102
Cu feet to cu metres	0.02831
Cu metres to cu feet	35.3147
Cu yards to cu metres	0.76455
Cu metres to cu yards	1.30795

Capacity

Pints to millilitres	589.26
Millilitres to pints	0.00175
Gallons to litres	4.546
Litres to gallons	0.2199

Weight

Ounces to grams	28.3495
Grams to ounces	0.03527
Pounds to grams	453.59
Grams to pounds	0.0022
Pounds to kilograms	0.45359
Kilograms to pounds	2.2046
Tons to kilograms	1016.04
Kilograms to tons	0.00098

Index

UNIVERSITY OF WOLVERHAMPTON
LEARNING RESOURCES